W.B Clarke

Remarks on the Sedimentary Formations of New South Wales

Illustrated by References to other Provinces of Australasia

W.B Clarke

Remarks on the Sedimentary Formations of New South Wales
Illustrated by References to other Provinces of Australasia

ISBN/EAN: 9783744750769

Printed in Europe, USA, Canada, Australia, Japan

Cover: Foto ©berggeist007 / pixelio.de

More available books at **www.hansebooks.com**

FIELD OF N.S. WALES.
I.A.F.R.S. &c.

Coal Measures. ▇ Granite.
Drift Pebbles. ▇ Slates. K. Oil-bearing deposits
portion not geologically examined.

REMARKS ON THE SEDIMENTARY FORMATIONS OF NEW SOUTH WALES.

ILLUSTRATED BY REFERENCES TO OTHER PROVINCES OF AUSTRALASIA.

BY THE

REV. W. B. CLARKE, M.A., F.R.S., F.G.S., F.R.G.S.,

MEMBER OF THE GEOLOGICAL SOCIETIES OF FRANCE AND AUSTRIA, VICE-PRESIDENT OF THE ROYAL SOCIETY OF NEW SOUTH WALES, &c., &c.

Fourth Edition,

Corrected up to 1878 and enlarged; with Appendices containing Lists of Fossils of New South Wales described by European Palæontologists.

SYDNEY: THOMAS RICHARDS, GOVERNMENT PRINTER.

1878.

R. ASTRON. SOC.

INTRODUCTORY NOTICE.

The *First* Edition of the following Memoir was written for and published in the "*Catalogue of the Natural and Industrial Products of New South Wales*," and forwarded to the Paris Universal Exhibition of 1867 by the New South Wales Exhibition Commissioners, at whose request it was undertaken.

It was re-printed at Melbourne in the "*Official Record of the Intercolonial Exhibition of Australasia*" in the same year, and was subsequently honored by being transferred to the pages of the "*American Journal of Science and Art.*"

The *Second* Edition was prepared for the "*Report of the Intercolonial Exhibition of* 1870, *at Sydney*," and was included with an Essay "*On the Progress of Gold Discovery in Australasia from* 1860 *to* 1871" (by the same author) in the work entitled, "*The Industrial Progress of New South Wales.*"

The *Third* Edition carried on the mention of Geological experiences as to the Sedimentary Formations of Australasia to the year 1875, and had reference to the Philadelphia International Exhibition of 1876—of the New South Wales Commission for which, and of those of the years abovenamed, the author had the honor of being a Member.

The present (or *Fourth*) Edition continues the Progress of Geological investigation up to date, and contains much fresh information. It is dedicated with great respect to the Congress of Geologists assembled at Paris in connection with the International Exhibition of this year.

Branthwaite, North Shore,
2 June 1878.   (*quo die octogenarius*)   W.B.C.

REMARKS ON THE SEDIMENTARY FORMATIONS OF NEW SOUTH WALES.

If we inspect the map of Australia we observe that the coasts of Victoria, New South Wales, and Queensland follow the general directions (with some irregularity) of the Cordillera, or elevated land separating the waters flowing directly to the coast from those which, draining the interior, disembogue to the south-west.

The Murray River receives some parts of its tributaries from the high lands of Victoria, and others from New South Wales; whilst the Darling and its tributaries collect the remainder of the supply from as far north as 25° s.

The Cordillera thus sweeps round in an irregular curve from w. to e. to the head of the Murray—and thence, northerly and north-easterly, to the head of the Condamine; trending north-westerly from that point to 21° s., whence it strikes to the north, terminating its course at Cape Melville, in 14° s., about the meridian of 144° 30′ e., which is that of Mount Alexander in Victoria.

The more westerly and southerly trend of drainage is represented by the Thomson and Barcoo Rivers, which carry off the waters of the Cordillera at the back of the Barrier Ranges to Spencer's Gulf. The meridian of the head of that Gulf is therefore the western limit of East Australia.

The Cordillera itself, described in part by Strzelecki in 1845, was traced by him through a considerable part of its diversified course (as understood by him), from the southern point of Tasmania to the parallel of 28°, in longitude 152°, but not further westward than 146°, on the parallel of Mount Alexander. It is, however, doubtful whether the range between this furthest western point and Wilson's Promontory, where he considers the chain to be cut off by the sea, forms anything more than a spur in that direction, though passing through Bass's Strait on to Tasmania.

But the extent of the Cordillera westerly, to its termination on the border of South Australia, is so well defined that there can be no question that the s.w. and w. extension has as true a character as any part of the northern prolongation. This may be geologically deduced from researches of the Geological Survey of Victoria. That province is limited, at its eastern corner, by a line joining Cape Howe and the head of the Murray, so that the boundary crosses very near the highest point of all Australia, which Strzelecki made 6,500 feet above the sea, but which subsequent observations have shown to be 7,175 feet. This correction rests on observations made by myself in 1852, and on a re-discussion of them in comparison with results obtained by

Professor Neumayer in 1862. On 8th May, 1852, I made the highest point of Kosciusco 4,077 feet above my then base, at 3,098 feet above the sea, which therefore came out 7,175 feet; and in February, 1863, Professor Neumayer wrote me word that he made the highest peak, in November, 1862, 7,176 feet. This makes Kosciusco's summit, above the crossing place of the Indi or Hume River, at Groggan's, 5,425 feet.

To the northwards, the 144th meridian limits very nearly all the high land of the East Coast to Cape Melville, whilst the 142nd meridian limits to the westward the basin of the Darling, including part of the drainage along the Thomson and Barcoo, from the head of the Flinders to where it passes into South Australia on the 141st meridian.

Thus, all this enormous drainage of western New South Wales and south-western Queensland is, as it were, bounded by ranges of high geological antiquity, the Grey and Barrier groups being of undoubted similar age to the mass of the eastern Cordillera.

It has long been known that the strike of the oldest Sedimentary rocks through the Cordillera, in Victoria, as well as in New South Wales, is generally meridional; so that in the former province the beds strike across the Cordillera, whilst in the latter they form various angles from parallelism with it to a transverse direction, as the chain doubles and winds irregularly in its course.

This is the experience of the Victorian Survey, and my own traverses across various points of the Cordillera in New South Wales and Victoria establish the fact of a normal meridional strike of the oldest strata. So distinct, indeed, is this characteristic, that the settlers in various parts of this Colony have been accustomed to trace the direction of north and south by the strike of the slates, and are often guided by it.

It sometimes happens that, owing to the high angle of dip, and the effect of denudation on the overlying formations, the Cordillera itself becomes in places almost knife-edged, so that in New South Wales it presents occasionally a watershed not more than nine paces in width; whilst in Mancero to the south, and in New England to the north, it spreads out in a plateau, on which eastern and western waters rise close together and sometimes overlap. These different features have a variable geological value as well as aspect; for, owing to the strike of the older rocks, the breadth of the Silurian formations, which, as in other countries, are repeated by recurring folds, may be more exposed in Victoria than it is in New South Wales; and owing to the curve of the Cordillera probably the same beds are traceable to the north which occur in the south; as, for example, the auriferous rocks of Omeo and Peak Downs, which are on the same meridian; and thus the meridional strike is exhibited along the north-east coast, where there are alternations of old rocks forming precipitous cliffs with low valleys and beaches separating those alternations.

Independently of this arrangement, the whole of the central area inside the eastern Cordillera has a trend to the south and west, so that the waters collected between 22° and 37° s., on the east of South Australia, find their way to the sea at the eastern corner of that province.

We might naturally assume that one order of deposits is to be expected throughout the Cordillera; but there is a singular exception. Whilst Marine deposits of Tertiary age are found along the west coast of Australia, and along the southern coast from Cape Leeuwin to Cape Howe, there are no known *Marine* Tertiaries in any part of the Coast of New South Wales and Queensland up to the Cape York Peninsula; and the reason of this may be, that, as indicated by phenomena before pointed out by me, but which on this occasion cannot be further dwelt upon, the eastern extension of Australia has been probably cut off by a general sinking, in accordance with the Barrier Reef theory of Mr. Darwin. This has some support from the fact that there is a repetition of Australian formations in the Louisiade Archipelago, New Caledonia, and New Zealand—in the latter of which occur abundant Tertiary deposits. The intervening ocean may therefore be supposed to cover either a great synclinal depression or a denuded series of folds; but, as shown in 1874 by the soundings from H.M.S. "Challenger," this depression is of enormous depth, in one sounding 2,625 fathoms having been reached.

Relatively speaking then, the Cordillera of the eastern coast has not been subject to the changes which introduced the relics of a Tertiary ocean, or they have been removed by subsequent sinking and denudation. At any rate, no evidence is known to me of *Marine* Tertiaries on the lands north of Cape Howe.

Another fact worthy of notice, as showing the probable ancient geological vicissitudes of Australia, is, that the great Carboniferous series which is so prominent in New South Wales and in parts of Queensland, but which is less distributed in Victoria, and there only partially and irregularly as to the portions still remaining, has been broken up and carried away, so as to have left the various members dislocated, ruined, and separated in such a way as to allow no clear view to be taken of the whole till all the various portions have been separately examined; and to the want of this personal examination on the part of certain Palæontologists and others who have never yet studied the Carboniferous formation of New South Wales, is to be attributed the pertinacity with which they have so long disputed facts attested by geologists in New South Wales who are familiar with that Colony and with Victoria also, but who are ignored by the closet-geologists of the latter.

In consequence of the absence of Marine Tertiary deposits in New South Wales, and the occurrence of a more complete series of the strata in the sections of the Carboniferous formation, there has arisen a difficulty in collating the gold-deposits with those of Victoria; and, in this respect, at present the Upper deposits in the former province have not been assigned with much precision to the epochs adapted by Mr. Selwyn for the latter. And it also follows that his view of the distinct ages of Pliocene auriferous and Miocene non-auriferous gravels cannot be tested in New South Wales, if indeed it has not already been tested by the actual discovery of gold in the so-called Miocene deposits themselves, as they occur in Victoria.

So far as is at present known, gold in Victoria is derived chiefly from the Lower Silurian formation; but researches conducted for me at H.M. Mint in Sydney prove that it exists in almost every distinctive rock of New South Wales. In this province the alluvial deposits are not so extensive as in Victoria; but this probably arises from the fact previously mentioned, of the strike being in Victoria transverse to the direction of the Cordillera, by which means the currents which distributed the drift had a wider area of gold-bearing materials to denude than in New South Wales, where, I conclude from numerous examples, the principal currents were to northward, so that in that province they would coincide with the direction of the Cordillera, and not accumulate the deposits in such low-lying extensive regions as those of the Murray Districts. The same objection would obtain on the supposition of gradual waste and accumulation from less powerful agency than that of a general rush of water. It is not however to be doubted, that there is an enormous amount of gold yet untouched in numerous places in New South Wales, not only in the quartz lodes (or reefs) but in gullies and plains where alluvial gold diggings will yet be discovered.

Dr. Duncan, in an elaborate paper on some of the Fossil Tertiary Corals of Australia ("*Proceedings of the Geological Society*," August, 1870), suggests the propriety of discarding the divisions into Pliocene, Miocene, and Eocene of the Australian Tertiaries, and of substituting the general term Kainozoic, since he considers them merely as successive deposits of one continuous epoch. But as proved by my own researches more than twenty years ago, much of the gold in New South Wales is derived from iron pyrites, in granite and in *beds* of Sedimentary origin consisting of siliceous matter cemented by iron derived from decomposed pyrites; whilst it has been shown by Aplin, Daintree, Hacket, Wilkinson, and others, that much gold in Victoria and Queensland is due to the intrusive agency of felstones, elvanites, and diorite. The dykes or reefs of quartz in the Silurians are therefore not, as once supposed, the exclusive sources of Australian gold. Nay, there is good reason to believe that the Carboniferous rocks are themselves impregnated, as in one remarkable instance on Peak Downs.*

In New Zealand gold sometimes occurs so mixed with siliceous particles as to constitute with them a golden sandstone.

The distinctive differences in material mineral wealth between Victoria and New South Wales are not altogether confined to gold or tin, which latter metal is well represented in New South Wales and Queensland; but coal, iron, and copper, and perhaps lead, prove together more than an equivalent of the great amount of gold in Victoria.

At the Universal Exhibition of Paris in 1855, the present writer exhibited a collection of rocks and fossils, illustrating the

---

* This example is thus referred to in a communication to me from Mr. Daintree, F.G.S., in a letter dated "Maryvale, North Kennedy, Jany. 22nd, 1870:—

"I believe if the Peak Downs District were carefully mapped, it would be incontestably proved that *payable* drift gold is there found in the Carboniferous conglomerates."

He then gives a section of the shaft and drive then being worked at the Springs, about 12 miles from Clermont, and adds:—"The miners use the Carboniferous sandstone, the Glossopteris bed at bottom, and take the cement several inches from its junction with the Glossopteris bed for their washdirt. The surface of the Glossopteris bed is unbroken, dips southerly at an angle of about 5°, and the cement lies conformably on it, and little patches of mud deposit in the cement, similar in appearance to the Glossopteris sediment, lie in the same plane as that bed, and I have no doubt the cement is conformable to the Glossopteris bed of the same period of deposit. Small fragments of Coal were taken from the adjoining shaft, and I have no doubt, with the necessary time given to the work, Carboniferous fossils may ultimately be found in the conglomerates themselves—so putting the matter beyond reach of dispute."

A similar instance of such an occurrence was examined by myself in the Coal Measure drift of Tallawang, in the county of Phillip, in the year 1875, and recognized as payable by C. S. Wilkinson, Esq., F.G.S., the present Geological Surveyor, in his report to the Minister of Mines, Dec., 1876, in which place there is mention of other notices by myself of like association. The localities are similar in geological structure; for almost in the words of Mr. Daintree, which Mr. Wilkinson never read, the latter says, "These conglomerates are associated with beds of sandstone and shale, containing *Glossopteris*, the fossil plant characteristic of our Coal Measures." ["*Annual Report for year* 1876," p. 173.]

I made a section of the deposits which I found resting on hard shales (probably Devonian) in which numerous shallow shafts have produced alluvial gold. The bottom of the beds above the base exhibited a brecciated fragmentary deposit, well seen a mile or two away, on the road to Cobborn—above which sandstones, flinty shale, coarse grits, the red shales of Mount Victoria and Blackheath occur; and, nearer the top, Vertebraria and Glossopteris and Charcoal are met with. One of the beds was of quartz-pebbles, cemented by ferruginous matter, precisely like many detrital fragments in other gold-fields, and specially resembling that above Govett's Leap, in which I obtained gold in 1863.

whole of the geological formations of Australia as then known, and these were enumerated in their stratigraphical order in the published catalogue. A few remarks on the various geological epochs, as they now represent themselves in New South Wales, with brief statements as to their connection with other portions of Australasia, may be all that is necessary on the present occasion, in addition to a comparison of the catalogue above referred to with the collection exhibited in Paris in 1878 by the Department of Mines, Sydney, and others, to show the progress of geological development in New South Wales during the last twenty-three years.

§ 1. So-called "Azoic" or "Metamorphic" Rocks.

There has not been sufficient evidence yet collected to show that these rocks extensively exist in Eastern Australia, although in Tasmania rocks of a doubtful class (and which may perhaps be only highly altered Lower Silurian) have been referred to them by Mr. Gould. The existence of gneissoid strata, and of schists of very ancient aspect, with occasional unfossiliferous limestones, are also well known in New South Wales, as at Cow Flat, near Bathurst; Cooma Hill, Mancero; Wagga Wagga; flanks of Mount Kosciusko, &c.; but it would be premature to place them, without doubt, under the present head. Mr. Daintree, however, describes them as the source of some gold in the Cape River and Gilbert Districts, to the North. Some of those mentioned under the "First Epoch" of Strzelecki, have, on close inspection, appeared to me to be merely the products of transmutation; nor is such an improbable result, seeing that in Australia some slates have been changed into granitic rocks. It is at least certain that such rocks generally occur in the immediate vicinity of granites, which latter frequently occupy large areas both in Mancero and in New England, as well as along the Cordillera, and in independent masses along the coast. In Western Australia, where an enormous region is occupied by granites, and the older formations are represented only by small patches of slates, whilst the granites themselves remain bare, these patches are found on the flanks of the granitic bosses and at extremely wide intervals; nor have I been able to detect among the numerous collections which have passed through my hands, any distinct evidence of any but doubtful examples of those foliated rocks which belong to the so-called Primary epoch. In Southern Australia, also, there does not appear to be any considerable amount of strata which could be referred to this epoch. Transmutation has, however, acted vigorously in New South Wales in all the older formations.

§ 2. Lower Palæozoic Rocks.—Lower and Upper Silurian.

Of these there are undoubted evidences in some limited districts of Tasmania and Queensland, whilst in Victoria and New South Wales considerable areas are occupied by them.

Western Australia has as yet not furnished any fossils of Silurian age; but, according to Mr. Y. L. Brown, Government Geologist, there are clay slates, schists, and other rocks which may be Silurian much transmuted, judging from their position and composition.

North Australia is in much the same condition, where no reliable geological surveys have been yet made.

Much valuable information was in 1864 collected by the Rev. Julian E. Tenison-Woods, and published at Adelaide, in which gold-bearing rocks were but slightly anticipated. Since then, the Northern Territory (assigned to South Australia) has exhibited gold-reefs in probably Silurian strata; and very recently a tract of several thousand square miles in extent, between the Victoria River and the Gulf of Carpentaria, along the Daly River and the central lines of communication by telegraph, has been reported as auriferous, and, I anticipate, will be found rich.

South Australia proper, according to Mr. Woods ("*Geology*," pp. 20, 21) has produced two Silurian fossils, *Cruziana cucurbita* and *Pentamerus oblongus*. The former occurs in Bolivia, and the latter in New South Wales.

Nothing lower than Siluro-Devonian, according to Mr. Etheridge (in "*Review of Mr. Daintree's Fossils*," Q.J.G.S., August, 1872), had up to that time been found in Queensland. But as elsewhere mentioned, I considered the Brisbane slates to be analogous with those of the Anderson Creek Gold-field in Victoria, both of which groups I examined personally *in situ*. The latter are held to be Upper Silurian.

In Tasmania, along the Gordon and Franklin Rivers, occur various Silurian fossils, some among which identical with those of New South Wales were noticed by me; but Mr. Gould considers others to be Lower Silurian. *This* formation evidently exists in that Colony, for in 1873 I received from Mr. T. Stevens, F.G.S., some Trilobite-sandstone from the western part of the Island, which Mr. Etheridge determined for me to contain *Phacops, Ogygia, and Calymene;* and to these Professor Bradley, of the U.S., to whom was forwarded for me by Professor J. D. Dana some of the rock, added *Conocephalites*, a true Lower Silurian fossil in America, Sweden, Bohemia, and Spain, a curious position for which in the last-named country is assigned in an interesting paper by Señor Casiano de Prado ("*Bull. Soc. Géol. de France*," 2de S., xvii, 516.)

Mr. Gould mentioned, in June, 1860, a *Calymene* at the base of the Eldon Range. I found that genus also in New South Wales in 1852. In Victoria Professor M'Coy has made a list of twenty-five Lower and fifty-three Upper Silurian fossils, including in the former twenty-three Hydroid Zoophytes, and another species belonging to the Upper formation. Of the Graptolitidæ only one is said to have been found in this Colony, and I presume that it is more likely to belong to the Upper Silurian than to the Lower, though towards the Victorian boundary, along the Deleget River, Lower Silurian rocks, according to some, are supposed to make their appearance.

New South Wales offers a more determined evidence of the existence of certain Silurian deposits, but singularly enough nothing has been positively shown of the existence of any fossils below the base of the Llandovery or the Middle Silurian, except in the case just mentioned.

To this epoch I referred fossils found by me in Mancero, in my Report of November, 1851, which was re-published in 1860; and it is satisfactory to find that the examination of a considerable amount of specimens by Prof. de Koninck of Liége, who kindly undertook the task of describing them, has resulted in a confirmation of my opinion. (*See Appendix XIV.*)

Summing up his review of sixty of these, he says—that they are in nearly equal divisions of the upper and lower beds of the Upper Silurian formation, and that they closely agree with the fossils of Europe and America; that the major portion of the former belongs to the Actinozarians and Crustaceans, and that the latter are nearly all Mollusca; and that none of the Graptolites noticed by Prof. M'Coy in 1861, and more recently by Mr. R. Etheridge, junr., from the Victorian strata, occur in the collection sent by me. And he concludes, as I have done, that at present the existence of fossil beds below the Middle Silurian has not yet been determined in New South Wales.

It is otherwise in Victoria, but it may be that some of the highly transmuted rocks of the south-west portion of New South Wales may yet furnish traces of greater antiquity when thoroughly examined. In the last Edition of this memoir, published in 1870, I mentioned the existence of certain Corals, Trilobites, &c., as determined for me in 1858 by the late Messrs. Salter and Lonsdale. (*See Appendix XVII.*)

The Mudstones of Yarralumla, with Encrinurus and Calymene; the Coralline and Pentamerus beds of Deleget and Colalamine; the Tentaculite and Halysites beds of Wellington and Cavan; and the Silverdale and Bowning beds with Calymene, Encrinurus, Beyrichia; and others with Illænus, Harpes, Bronteus; Brachiopoda, including Strophodonta; and Radiata, embracing Star-fishes— point to the existence of at least the Upper Silurian formation on both flanks of the southern part of the Cordillera. There are also numerous corals, included in the list given by me in the "*Southern Gold Fields*" (p. 285), which also confirm the same determination; and it may be added that the above, and other fossils of this age mentioned by me elsewhere, have been examined by Palæontologists of eminence in Europe. Such are the genera Favosites, Cœnites, Ptychophyllum, Calamopora, Syringopora, Emmonsia, Alveolites, Cystophyllum, &c. These, perhaps, might not alone satisfy a doubt, but with them occurs Receptaculites; since 1858, when these were determined, I have detected Halysites, which may settle the question as to Upper Silurian. Wenlock beds seem to be well developed on the Deleget River.

Professor De Koninck is not in antagonism with these geologists, but in the fresh series of my fossils he found among the trilobites Staurocephalus, Cromus, Proetus, and Lichas, in addition to Calymene, Encrinurus, Illænus, Harpes, and Bronteus before announced by myself. (See Edition in 1870, p. 6; and ".*Southern Gold Fields*, 1860," p. 286.)

In a paper published by the learned Professor, in the "*Mémoires de la Société Royale de Liége*," 2de Serie, t. vi., 1876, dedicated to the Silurian and Devonian species of N.S.W., forwarded to him for his examination and description by myself, he gives those, as detailed in Appendices *XIV* and *XV*, which I have thus above epitomized. The description is given in a separate form, with carefully executed figures, under the heading of "*Recherches sur les Fossiles Paléozoiques de la Nouvelle-Galles du Sud*," in which, as will be noticed under the next section, are included the Devonian fossils.

§ 3. MIDDLE PALÆOZOIC ROCKS.

The late Mr. Jukes desired the term Devonian to be eliminated, referring the so-called beds to the bottom of the Carboniferous formation; but geologists have not generally accepted that proposal. The series of shells, corals, &c., from the Murrumbidgee, which I submitted in 1858 to Messrs. Salter and Lonsdale, through Sir R. I. Murchison, Bart.,* excited doubts as to their belonging to any but Silurian and Carboniferous deposits. Among these were Phanerotinus, Loxonema, Atrypa *reticularis*, Orthis *resupinata*, Murchisonia, Strophomena, and Spirifera of various species.

Mr. Salter's Report to me was as follows: "These fossils are of a mixed character, many being of unquestionable Silurian age, and others having all the aspect of Carboniferous and Devonian

---

* See Murchison's "*Siluria*," 3d ed., p. 226, and 4th ed., p. 276 and p. 462.

fossils. It will not be so easy to predicate those of Devonian type, as there is much similarity between fossils of that age and those of either of the other systems, the Lower Devonian species being very like Silurian, and the Upper Carboniferous ones. But if none of the fossils came from Carboniferous beds, then there must certainly be Devonian forms mixed with Upper Silurian."

Mr. Morris contributed, in 1845, a paper to Strzelecki's work of that year, in which he says: "The Palæozoic series of Australia and Tasmania may be regarded as partly the equivalent of the Devonian and Carboniferous systems of other countries." (*See Appendix VII.*)

In September, 1859, I addressed a letter to Mons. Delesse, which he communicated to the Geological Society of France, in November, and in the report of the meeting (*Bull.* xvii., p. 17) I find I expressed myself cautiously as follows:—" Le devonien et le permien sont probables sur quelques points mais peu distincts."

In 1861 (*Cat. Vict. Exh.*) Professor M'Coy stated that "there had as yet been no exact identifications to prove the existence in Australia of the Intermediate Middle Palæozoic or Devonian formation." And as recently as 1866, Vicomte d'Archiac (" *Géologie et Paléontologie,*" p. 468), writes thus : " Le développement des séries Siluriennes et Carbonifères dans l'Australie doit y faire soupçonner entre elles un représentant de celle qui vient de nous occuper ; mais il ne semble pas qu'elle y ait encore été bien charactérisée par ses fossiles."

About the same time Professor M'Coy ("*Exhibition Essays of 1866-7*") mentioned that the limestones of Buchan, in Gippsland, contained " characteristic corals, *Placodermatous* fish and abundance of *Spirifera lævicostata*, perfectly identical with specimens from the European Devonian limestones of the Eifel." In the Melbourne " *Official Record of the Exhibitions of* 1872-3," the addition of some other places in Gippsland (unnamed) and of Mount Gibbo, is introduced ; and in 1874 there was included in the " *Progress Report of the Geological Survey of Victoria,*" a list of fossils of the most characteristic common types, drawn up by Professor M'Coy, which, under the head of Devonian, includes the following : *Favosites* (two species), *Spirifera lævicostata, Grammysia* (n. sp.), *Orthonota* (n. s.), *Asterolepis* (plates allied to). In 1847 the same skilful Palæontologist noticed some striking resemblances to Devonian fossils in a few of the large collection I sent in to the Woodwardian Museum at Cambridge ; and Professor De Koninck, also in 1847 (" *Recherches sur Animaux Fossiles*") records Sp. *Murchisonianus*, a Devonian fossil from Tasmania.

In order to test the existence of a wide-spread Devonian series in New South Wales, I requested (as stated elsewhere) my friend Professor De Koninck to undertake the examination of a collection of 1,000 Palæozoic fossils, comprising the Upper, Middle, and Lower Palæozoic formations as they exist here, and in his account of the Devonian, received since the last edition of this memoir, he remarks in his "*Résumé Géologique*" (op. cit. p. 133), after giving the fullest assurance of all possible accuracy:

"Des quatrevingt-une espèces observées en y comprenant
"un spongiaire nouveau, mais non décrit à cause de l'impos-
"sibilité d'en définir le genre, ainsi qu'une tige de *Rhodocrinus*,
"il n'y en a que cinq qui puissent être considérées avec certitude,
"comme provenant des assises Dévoniennes Supérieures.
"Ce sont,—

"*Strophalosia productoïdes*, Murchison.
"*Chonetes coronata*, Conrad.
"*Rhynchonella pleurodon*, Phillips.
"*Spirifer disjunctus*, Sowerby.
"*Aviculopecten Clarkei*, L.-G. de Koninck.

"Toutes les autres, ou du moins le plus grand nombre et
"principalement celles qui se trouvent dans le calcaire noir des
"environs de Yass, appartiennent à un horizon géologique un
"peu inférieur à celui qui a fourni les espèces que je viens
"de signaler, mais cependant plus récent que celui qui est si
"bien caractérisé par la présence de la *Calceola sandalina*,
"Lamarck, dont je n'ai pas rencontré de traces, pas plus que
"des *Trilobites* qui l'accompagnent ordinairement. Parmi ces
"quatre-vingt-une espèces, trente sont nouvelles pour la science
"et ne sont connues qu'en Australie, mais il est à remarquer
"qu'à l'exception de quatre d'entre elles, toutes ont leurs
"analogues en Europe et en Amérique. Ces quatre espèces
"sont,—

"*Archæocyathus? Clarkei.*
"*Billingsia alveolaris.*   } L.-G. de Koninck."
"*Niso? Darwinii.*
"*Mitchellia striatula.*

The author goes on to say, that the first of this group of four appears in Australia to occupy the place which in certain beds in Europe, and very particularly in Belgium, is held by *Receptaculites Neptuni*, Defrance, which as the other belongs to the order of sponges.

As the collections under review were made in part before 1850, having been packed some years before they were sent for examination to my highly honored friend, much correspondence has taken place between us and as I have, since the specimens were received by him, made numerous explorations, and during these have very extensively collected from the region along the Yass and Murrumbidgee Rivers (in continuation of my earlier researches), and have had the opportunity of being accompanied in 1876 by Mr. Jenkins, of Yass, whose acquaintance with the palæontological treasures of that neighbourhood is very great,—it has been my good fortune to find the missing *Calceola* and numerous *Trilobites* alluded to in the preceding extracts from De'Koninck's admirable "*Recherches*." At present I am unable to submit these Yass fossils for description.

In addition to *Calceola* (which occurs also at Mount Frome, County Phillip) I have been also able to satisfy my friend that *Receptaculites Neptuni* also exists in New South Wales, as well as *R. Australis*, which was sent by me to the late Mr. Salter and figured by him in the "*Decades of Organic Remains of the Geological Survey of Canada*," in comparison with *R. Occidentalis* syn. of *R. Neptuni* (See *Decade* I, pp. 45-47, pl. x, figs. 1-10). (*See Appendix XVII before cited.*)

It is true that Mr. Salter regards the *R. Australis* as Upper Silurian, and rightly associates it with Tentaculites, Favosites, Pentamerus, Orthoceras, Trochonema, Rhynconella, &c., which I discovered in the south-western district of Maneero in 1851-2, and also that *R. Neptuni* came from between Wellington and Molong; and that the actual limits of the Upper Silurian and Devonian formations have not yet been accurately defined. But when we find such a genus as *Niso* represented in a Palæozoic formation, as is the case with the Devonian of New South Wales, and notice how frequently of late palæontologists have been obliged to admit the occurrence of genera and sometimes of species of acknowledged younger formations in those of more ancient date—as anticipatory of future existences—it may be well believed that *Receptaculites* may be generically known to such double relations as the Silurian and Devonian. (*See infra*, p. 97.)

"C'est pour la première fois," says De Koninck," que la présence du *Niso* est signalée dans les terrains Paléozoiques, et il faut remonter jusqu 'au terrain Tertiaire pour en retrouver de nouveau les traces; cependant mon savant et excellent ami M. Nyst, sans contredit un des meilleurs conchyliologistes de l'époque, que j'ai consulté à cet égard, croit pouvoir déclarer avec moi, *qu'il n'existe pas une différence suffisante entre les caractères généraux de l'espèce Dévonienne et ceux de l'espèce Tertiaire pour ne pas considérer l'une et l'autre comme génériquement identiques.*"

I may add here, that some years since, I sent to H. M. Jenkins, Esq., F.G.S. (when Curator of the Geological Society), a species which he considered to be a *Lepralia*, which was bedded in the limestone of Cavan on the Murrumbidgee and in the same geological district as that now discussed. M. De Koninck also notices, in addition to *Niso*, the occurrence of *Mitchellia* in the midst of Marine shells, as another striking anomaly in the Devonian *fauna*, compared with that of Europe, concluding in these words,—" Elle n'est certes pas suffisante pour empêcher de considérer l'une et l'autre comme contemporaines et produites dans des circonstances sinon tout à fait identiques, au moins très-analogues." (p. 135).

It will be seen on perusing the lists of Devonian fossils, that De Koninck includes those I referred to in the second Edition (1870) of this memoir, from Yass, Mount Lambie, and Moruya River, and which are in part identical with the Mount Wyatt shells in Queensland.

These latter are mostly Brachiopods, and I have collected them during my different journeys of several years from the western boundary of the Carboniferous formation (underlying it *in situ*), and occasionally from a scattered over-lying drift, ranging for nearly 260 miles of direct distance (included between 36° south on the Moruya, to nearly 32° south). The principal of these particular Brachiopods are—*Rhynconella pleurodon*, *R. pugnus*, *Spirifer disjunctus*, *S. Yassensis*, *Orthidæ*, *Productæ*, &c. They occur *in situ* between the slaty rocks of Sofala and the overlying Carboniferous beds on the Turon; south of Moruya River; near Mullamuddy on the Cudgegong River; at Cudgegong Creek; in the deep defiles of the Upper Colo River; at Brucedale and Bathurst; and in other places. Mr. C. S. Wilkinson, with whom I visited the locality in 1875, found them under interesting circumstances occurring in a great synclinal curve, from nearly the summits of Mount Lambie and Mount Walker (with considerable dips), and explaining the sources from which the loose pebbles collected by me at Bowenfells some years since were probably derived. From the occurrence of different fossils in the pebbles, it is certain that many strata of the Devonian formation must have been broken up, and it seems that similar beds have undergone the same process in other countries, for I well remember picking up in 1829, in the "Platz" of Coblentz, on the Rhine, a similar drift pebble, of just such rock as that in question, containing a Brachiopod of like age.

During some recent explorations in the north-west of this Colony, I became satisfied as to the widely-spread extent of the Devonian series, of which more evidence will be elicited hereafter, the data for which are already sufficient, but there is no room to introduce them on this occasion.

I may however mention now, that amongst the specimens collected by me in the neighbourhood of Yass in 1876, I find a portion of an Ichthyodorulite, which I believe to be Devonian, and that in March, 1878, Mr. C. S. Wilkinson sent me for comparison a specimen of fossiliferous limestone, which I find also came from the Murrumbidgee not from Yass, and which contains a plate of a Coccosteus, of a triangular shape, studded with tubercules of the same form as those on a plate of M'Coy's *C. trigonaspis*, but somewhat different on the whole from his figure, (See "*British Palæozoic Fossils,*" Pl. 2. C. Fig. 6 e.) It is attached to a portion of bone, and is in good preservation and in the midst of fragments of other fossils, the matrix being apparently the same as the Yarradong or Cavan limestone. It was found by Mr. Hume.

Tasmania has at present furnished no well-established proof of the distinct existence of Devonian rocks. But it is a fair inference, first suggested by the late Mr. Salter, that the broad-winged Spirifers common there in the Palæozoic beds imply the probable occurrence. Mr. Jukes and Mr. Gould both repeated the inference. Mr. Darwin and Mr. Selwyn agree that some of the Tasmanian fossils "occur in the Silurian, Devonian, and Carboniferous strata of Europe." This is nearly all that is known respecting their position.

Western Australia, according to Mr. Brown's Report, adds nothing to the history of the Middle Palæozoics; but Mr. H. Gregory indicated on his map and in his report the existence of Devonian rocks near York and in other parts of that Colony. Having examined the rocks so indicated, I can only state my belief that they have no pretension to any such antiquity, and are probably mere collections of loose granitic matter and other drift cemented by ferruginous paste, which has since become transmuted into concretionary nodules and hæmatite. There are also pebbles of trap, much decomposed, in the so-called Devonian. They may perhaps be more properly considered as representing the *laterite* of India.

Queensland, on the other hand, according to Daintree's notes, exhibits a stretch of Devonians extending through *ten* degrees of latitude. Not the least interesting facts are that the Tin Mines of Queensland (as well as those of New South Wales) occur in granites of Devonian age.

At Gympie, on the River Mary, rich gold-bearing quartz-reefs occur in transmuted slates and other tilted beds, which are composed of detrited dioritic matter and brecciated deposits in which are abundance of fossils of doubtful aspect, and these I before referred to some part of the Carboniferous formation. Mr. Etheridge considers and has described the fossils as Devonian. They certainly have much in common with the Devonian beds of North Germany and Belgium, described by Sedgwick and Murchison, as I stated in the Second Edition, p. 10. It is right, however, to remark that Professor M'Coy does not adopt this determination, considering the rocks to be younger.*

Whatever be the age of the Gympie beds, in rocks of apparently the same age in Queensland there is a vast amount of mineral

---

* The following are the grounds upon which I ventured an opinion as to their Carboniferous age in 1871 (" *Progress of Gold Discovery from* 1860 *to* 1871," pp. 5–7) :—

Notes on Mr. Hacket's Collection of Rocks and Fossils from the Gympie Gold-field.

1. This collection comprises two series of rocks, the one Sedimentary, the other intrusive.

2. The latter consists of varieties of the greenstone group of the Plutonic formations.

3. The former embraces several kinds of rock. Among them are some so completely free from transmutation as to exhibit the characters of ordinary schist, sandstone, and breccia; others appear to have been derived from volcanic ash of the dioritic type, and have been, since their deposit, altered by intrusive agency so as to put on the resemblance of diorite or greenstone, and as such they have by some been classified.

4. The presence of fossils serves, however, to illustrate their conditions as ash-beds deposited in an ocean troubled by contemporaneous or subsequent igneous action, which, after the consolidation (in part) of the strata containing organic bodies, became changed by the new eruption. A considerable portion of the Gympie Gold-field has thus become a metamorphic area.

5. Such phenomena are by no means rare in Australia. Bedded, as well as intersecting, basalt occurs largely in the Illawarra Carboniferous district of New South Wales, whilst, in the western border of that Colony (as about Wellington) greenstone is exhibited in a similar connection with Upper Silurian Strata. At Waimalee (Prospect Hill, near Parramatta) an old diorite, precisely like that of Bople, to the eastward of the Mary River, has furnished a matrix for the plant beds of the Wianamatta Rocks, the highest in the New South Wales series of Sedimentary deposits ; and these have been subsequently transmuted by younger igneous rocks that pierce and overflow them.

6. The whole of the Sedimentary deposits in Mr. Hacket's collection betray the effects of contemporaneous independent forces. The purple schistose rock contains, besides an occasional fossil, fragments of igneous products, and some segregated quartz; and the gray and greenish fine-grained stone, derived from dioritic detritus, contains frequently much lime, many imperfect squeezed fossils, with a portion of some drifted matter. Patches of the purple schist occur in the green rock ; and in the brecciated beds composed of fragmentary materials (the result of violence and subsequent consolidation in a state of repose), chemical action has produced segregations of quartz which simulate true quartz veins.

7. It is to be presumed that the fissures in the strata which are now filled in with auriferous and cupriferous quartz were formed at a later period. A considerable time must have elapsed, for many of the fossils are themselves changed or partly obliterated, and are traceable only by the glistening cleavage of calcareous sections.

8. Mr. Hacket has marked one variety of rock Schalstein, and it certainly agrees with the definition of that species, inasmuch as it is laminated with thin partings or coatings of calc-spar. Now this is a very common occurrence in parts of Germany where greenstone is also present, and where the age of the rocks is Devonian. Schalstein is truly a derivative and not an independent product, and therefore must be included with the other transmuted deposits. This rock exhibits at Gympie an exact resemblance to its namesake

wealth besides gold, as ores of copper, iron, tin, lead, antimony, mercury, &c. The work entitled "*Notes on the Geology of Queensland, with the Appendix of Animal Fossils: By R. Etheridge, Esq., F.R.S., F.G.S., Palæontologist to the Geological Survey of Great* on the Lahn, in Germany, where also are traces of copper ores and jasperised schists, as at Gympie. Mr. Hacket's excellent map of the Gympie Goldfields should be studied in connection with the valuable memoirs of Sedgwick and Murchison, in the "*Transactions of the Geological Society of London*," 2nd series, vol. 6 :—"*On the Older Deposits of North Germany and Belgium.*"

9. There is another probable connection between these Gympie beds and those just referred to. At any rate, so far as the fossils go, they lead to the conclusion that they are not older than Devonian, and may be Upper Palæozoic. The principal fossils capable of indication are Nucula, Fenestella, Solarium, Spirifera, Orthonota, Edmondia, Stenopora, and Producta, which last alone proves some of the beds to be not above the Upper and not below the Middle Palæozoic periods.

10. If this view is maintained, then we have evidence at Gympie, which is well supported elsewhere in Queensland and in parts of New South Wales, that auriferous quartz-reefs occur in rocks younger than Silurian; and we have there also an additional proof of the influence of greenstone in the production of gold deposits. The fact was many years ago pointed out by myself and by Mr. Odernheimer in relation to the Peel River Gold-field, and it has since then been extensively confirmed in the Thames River Gold-field in New Zealand.

11. In Mr. Aplin's report of July 21, 1869, mention is made of the resemblance of fossils in calcareous grits at Canal Creek to those in the "diorite slates" at Gympie. The beds there are said to form "*a narrow band between the greenstone area and the river.*" In these strata, though placed under the head of "Silurian beds," the principal fossils are Spirifera and Producta. It is more than doubtful whether Producta has ever been found in the Silurian formations, and it is held to be the most distinct of all fossiliferous tests of the epoch to which it is confined. So far as is known, it belongs to the Upper and Middle Palæozoic, and ranges only from Permian to Devonian formations. Assuming this limit for Canal Creek and Gympie, it becomes certain that beds of the age to which the fossils belong have a wide range in Southern Queensland, and this is the case in Northern Queensland also. Evidence will one day be produced to prove the occurrence of gold in the Upper Palæozoic formations in other localities. Nay, Mr. Daintree has given me his reasons for believing that it so occurs on Peak Downs. (See quotation in note at p. 9.)

The association of greenstone rocks with beds containing the fossils indicated, will form a guide for prospectors in fresh districts of the Colony.

Too much importance cannot therefore be given to the establishment of the fact to which the researches of Messrs. Hacket, Daintree, Aplin, and Ulrich have contributed, that igneous rocks of a certain class are the surest indications of gold in Queensland.

Mr Etheridge figures the following species as Devonian from Gympie:— ? *Aviculopecten limæformis*; ? *A. imbricatus*; *A. multiradiatus*; *Spirifera dubia*; *S. undifera*; *Strophomena rhomböidalis*, var. *analoga*; *Pleurotomaria carinata*; *Euomphalus*; *Fenestella fossula*; *Ceriopora* ? *lara* (Daintree's "*Notes on the Geology of Queensland*," Q.J.G.S., Augt., 1872, pp. 326, 333); others are mentioned as *Edmondia concentrica*; *Productus cora*; *Spirifera bisulcata*; *S. undulata*, &c.

De Koninck, 1877, considers the series to be Carboniferous, naming some of those given above as younger than Devonian. (*See Appendix XVI.* "*C.*")

Britain, and W. *Carruthers, Esq.*, *F.R.S.*, *F.G.S.*, *Keeper of the Botanical Department of the British Museum*," is an invaluable document, and deserves consultation (Q. J. G. S., vol. xxviii, pp. 271, 360.) The map, especially the large independent edition, and the plates and other illustrations, are highly useful.

It is interesting to find Dr. Hector stating at the beginning of 1875 that 2,000 specimens of Lower Devonian or Upper Silurian fossils have been obtained from the north-west district of the South Island of New Zealand ("*Ninth Annual Report of the Colonial Museum*, 1874.") And equally interesting is it to know that New Caledonia also holds out hope of contribution to the Middle and Lower Palæozoic faunas, as in the Isle Ducos, Leptæna, Spirifera, Orthis, &c., occur with *rolled Brachiopods* of the same character as those of the "Gulf" on the Turon River of this Colony. ("*Annales des Mines*," tome xii, p. 54, 1867.) Monsieur M. P. Fischer is disposed to assign them to the *Devonian* period ("*Bulletin de la Soc. Géol. de France*," 18 Mar., 1867).}

In further reference as to New South Wales, it may be well to mention that there seems to be in parts of the Western Districts an exhibition of rocks which resemble in various ways the conglomerates of the "Old Red Sandstone" of Europe; such overlie the Marine Upper Silurian beds in the neighbourhood of Wellington and elsewhere, and are known to contain Lepidodendra. These may be well studied in the ravine of Curragh Creek, where they overlie the Favosites beds of Jew's Creek, &c. They form ranges of considerable extent and of prominent features, and stretch, according to my observations, to the Coutombals and in patches as far as the Lachlan. The occurrence of a peculiar species of Lepidodendron in three of the Colonies, New South Wales, Queensland, and Victoria, has given rise to much controversy as to the age of the rocks in which it occurs. It has been long known to me, but it is only recently that it has been found by me in widely distant localities, sometimes solitary, at others in beds in which other plants of similar age occur.

It seems indeed as if every individual discovery in the Geology of this Colony had a history or literature of its own.

In June, 1851, Professor M'Coy wrote to me from Cambridge respecting the first Lepidodendron he had seen from Australia, and which I had forwarded by the late Rear-Admiral King to Professor Sedgwick, and stated it to be *L. tetragonum* of the English Coal-fields.

The late Mr. Salter, in his letter to me of May 9, 1859, said, however, that the genus was *not Lepidodendron.*

In November, 1863, Sir C. Bunbury wrote to Professor R. Jones respecting a collection of Australian fossil plants, including the above species, sent home by me and now in the Museum of the Geological Society, where they were inspected by him at my request, and noticed one (*the* one) which he considered to be very like *L. tetragonum*.

During the last few years I have collected or received this plant from a variety of localities in New South Wales and Queensland, and from the latter Colony it was also brought in abundance by Mr. Daintree. Mr. Carruthers, who has given its description fully in the paper before alluded to (Q.J.G.S., Aug., 1872), has assigned to it the name of a species described by Unger, viz., *Lepidodendron nothum*.

The extent of territory from which my specimens have been collected embraces a direct distance of more than 1,100 miles (English) between 19° s. and 35° s. (of course at intervals only), from which we may infer the importance of its discovery in any new locality, as establishing the existence of a portion of the Devonian series to which it has been finally assigned.

It was satisfactory to be able to recognize this plant in January, 1875, in a creek near Rydal, on a spur of the Mount Lambie Range, where the Devonian Brachiopoda occur, and to be able to direct Mr. Wilkinson to the locality where he found his five additional specimens, which certainly established the position *in situ* of the species near that locality.

But it was not until subsequent visits in 1876 and 1877 that I was enabled to detect the plant actually *in situ*, and which enabled me to ascertain the proper position of its habitat, which is considerably below the level of the Brachiopod sandstone of Mount Lambie, and on a spur of that range overlooking Solitary Creek. As a guide to future explorers, I left a "broad arrow" mark on the fence nearest the spot.

Professor M'Coy adheres to his first opinion that the plant is not *L. nothum* (See "*Coal Report*," and *Decade I*, "*Prodomus of the Geology of Victoria*,"—pl. ix), and calls it *L. Australe*.

Writing in 1861, the learned Professor proves that there is no mistake about the identity of the plant in question ; for he says, a specimen of it, still I believe in the Melbourne Museum, is of the same species as *the only Palæozoic Coal-plant* ever collected in New South Wales, and which was sent to him about twelve years ago for "determination during the controversy as to the age of the plant-beds of the Newcastle N.S.W. beds." This mistake as to *date* is of no importance, as it is rectified by my previous quotation from Mr. M'Coy's letter, and I only refer to it to show, which is due to himself, that we are treating of the same plant.

But there may be found many, and there are already known in New South Wales several distinct species of Lepidodendron and its allies, in the Lower Coal-beds along the Karua Basin and elsewhere; and although in the Upper seams they may not be known, we have the testimony of Sir Thomas Mitchell, on the authority of the late Mr. Lonsdale ("*Parliamentary Papers*, 14 *January*, 1852"), that a Lepidodendron was found "in sandstone" between Windsor and Parramatta, that "sandstone with impression of Lepidodendron was also found over granite near Cox's River," and I possess a cast of a Lepidodendron (verified as such by Mr. R. Etheridge, junr.) in sandstone also, from the banks of the Warragamba, between the junction of the Wollondilly with Cox's River and the junction of the Warragamba with the Nepean, being one in a Devonian area, and the others in the Wianamatta Basin. And if it be argued that they all came from one source, and were drift fragments (which may be or not be the case) still, as we discover elsewhere in Australia that the drift Lepidodendra are almost all found not far from the parent beds, the inference would naturally be that these had not drifted far to the localities now pointed out.

Mr. M'Millan himself told me in Melbourne in 1860 that the Lepidodendron found by him, and which was I presume the one I saw in the Museum, and which is figured by Professor M'Coy, was picked up from the surface and first used to keep a door open or shut in a store at Melbourne. But if it was not taken from its bed it is of no more value than any of those scattered about the surface of Australia elsewhere. But the *inferential value* is the same in all the cases.

The subject, however, with which we are now concerned need not depend on such data.

Professor M'Coy states that "*the sandstone* containing the present species in Victoria has been found by Mr. Howitt over a large extent of Gippsland to lie always unconformably on the upturned edges of the true Devonian rocks"; * and Mr. Selwyn mentions other specimens of Lepidodendron from the Avon.

These admissions are worth very much to any controversy I may have had with the able and skilled Palæontologist of Victoria, whose judgment as to genus or species I have no pretensions to

---

* I now recollect that Professor M'Coy has admitted this fact, "as to the specimen he alludes to in the Melbourne Museum, the Government Geologist can testify that on first seeing it some years ago in a *store at Melbourne*, I at once characterised it to him as the most important palæontological specimen ever found in the Colony," &c. ["*Commentary*" *read before the Roy. Soc., Melbourne*, 25th June, 1860.] But in 1861 the Professor had not apparently realised the value of the find, for he says in the "*Catalogue of the Victorian Exhibition*, 1861," p. 164: "Having as yet seen no distinct identifications to prove the existence in Australia of the intermediate Middle Palæozoic or Devonian formations;" but in 1857 he speculated on the occurrence of the Carboniferous formation all the way from the Avon into N. S. Wales. [*Evidence before Parliament*, "18th August, 1857." "*Progress Report on Coal Fields*," see *infra*.]

dispute, but who, like all the world, must necessarily be amenable to facts and logic; I therefore forego all comments on *L. nothum* and *L. Australe*, and leave the decision where it should rest—in the hands of Palæontologists.

Two points, however, remain for remark on the Devonian.

Gratified as I always am (when I consider that I have never had an hour's assistance in the field from any individual during my thirty-nine years of geological labour to profit by the collected intelligence and the host of well-skilled physicists that help to make geology what it ought to be) to peruse the recitals of the field-geologists of Victoria, I have had especial pleasure in reading Mr. Howitt's description of the Devonian area in Gippsland, and of his researches on the Snowy River and other localities in Victoria which I visited and examined in 1851.

In those days the Devonians were unknown to Australian explorers, and even so late as 1863 only a suspicion existed as to the actual relation of the strata about Mount Tambo and other places in Gippsland.

The limestones of Buchan and Bindi had not been as now correlated with Devonian, and the Director of the Geological Survey of Victoria himself, in 1866, held the opinion that the arrangement was such that the limestones were *Upper Palæozoic*. It is not surprising therefore, that in the patch of fossiliferous rock in a narrow gully at the very head of the Murray River which I discovered in December, 1851, I should have considered it "not younger than the base of the Lower Carboniferous"—which justifies the remark I made in the last Edition, that if Professor M'Coy was "*right*" in determining the Devonian to be the epoch of these Gippsland limestones, I could not be far *wrong*, especially considering that my discovery of fossiliferous rocks was made in a great hurry for want of time to institute careful search, during a journey in unpleasant weather, after lying all night on the bare ground, on the upper slope of the Great Dividing Range, at Kurnoolee (Native Dog Creek), on my way from Moambah to Omeo. In 1870 the remark was made by me that "*in* 1851" I held Mr. Selwyn's view, but I had no intention of disputing subsequent results; I do not think I was justly dealt with when I was reproved for so doing, especially as in 1870 I had written in the following words to the Editor of the "*Sydney Herald*" (in reply to a statement in the "*Ovens and Murray Advertiser*" respecting an alleged discovery of an *Ichthyosaurus* at the head of the Indi or Murray River):—"On 16 December, 1851, I visited the locality and found Marine fossils there of an age not younger than the Lowest Carboniferous rocks of N. S. Wales. The Victorian geologists consider them Devonian."

Mr. Selwyn reported plants from Mount Tambo, but Mr. Howitt ["*On Devonian Rocks of North Gipps Land. Progress Report*," No. III, p. 233] says that in 1876 he could not find any. That locality is not very far from the one visited by me. I certainly saw no Lepidodendron near Tambo or at the head of the Indi, but fossil wood occurred at Tambo Bluff, where it had also been found by Mr. John Wilkinson and Mr. Tyers, then engaged in their survey; before 1850, and in a letter written in that year, about twelve months before I visited Omeo, the latter spoke of the Bindi and Buchan limestones as mountain limestone, which I mention merely to show that others beside myself had at that time the same impression respecting the Sedimentary deposits of the vicinity.

The subsequent investigations by Mr. Howitt proved that the deposits in question were Middle Devonian.

My object in referring to this otherwise unimportant matter has been to explain the last passage in the note below (quoted from the last Edition of this memoir, as it appears in "*Mines and Mineral Statistics of New South Wales*," 1875, p. 161, and which was accidentally left incomplete.*

These references and quotations bear upon the possible relation of Lepidodendron to the Devonian Marine fossils in the localities mentioned.

Mr. Howitt's Report is very valuable on many accounts. He has made out with considerable precision the actual sequence of the Devonian beds over a large area, and places them not only as they should be above the Silurian, but as having once

---

* NOTE.—In Queensland, the Burnet Range, the Mount Wyatt District, and tracts about the Bowen Gold Field and Burdekin (on which river limestones with fossils occur), are strewn with spoils of a formation which Mr. Daintree calls Devonian. From the former locality I have had many collections, and among them all I find Productus in alliance with Trilobites which appear to be older than Carboniferous. On the western flanks of the Cordillera near Yass, and on the eastern along the Shoalhaven River, and again near the Hanging Rock, New South Wales presents numerous bands of limestone full of such fossils; but it is doubtful at present whether these lie on the horizon of the Devonian, or whether they belong to some portion of the Upper Silurian. As these beds appear to range all through the country on a nearly meridional strike on both sides of the Cordillera, they are traceable in widely different places; and it may eventually be determined, that though in close contact, there is really a distinction of formations only to be detected by accurate survey. So far as Lepidodendron is concerned, that plant occurs in some places in association with beds that are decidedly younger than any called Devonian, near Pallal on the Horton River, and on the Manilla River in Liverpool Plains, and in the gold-drift of the Turon River, which has been derived from beds of transmuted sandstone belonging to the Coal-beds at the head of the river. It occurs thus on Dangera Creek, Yalwal. Near Wellington, also, Lepidodendron has been found in hardened rock of similar origin. At Canoona Gold Field, in Queensland, Lepidodendron occurs in hardened shales; and at Goonoo Goonoo, on the

occupied a very much wider area than at present over what is now called North Gippsland, the Upper Devonian having occupied the whole of it. He seems to have assigned the Snowy River Porphyries in some localities to a position between Upper Silurian and Middle Devonian. The Porphyries are considered *generally* as Lower Devonian, resting as they do on Lower Palæozoics or Granite.

Under the circumstances detailed, there was no great heresy in considering the deposits hastily observed, as 1 then supposed them to be, as Lower Carboniferous, which was the oldest Sedimentary deposit then known with any certainty; and Mr. Howitt, in 1875, admitted that there is in Gippsland a passage from the Upper Devonian into the Carboniferous beds. "We find," he says, "that the materials of which the groups are composed are threefold—coarse conglomerates, sandstones, and shales with occasional beds resembling 'cornstones' in their calcareous character. No such unconformity is probable between the Upper Devonian and Carboniferous as between the former and the Middle Devonian." (p. 237.)

The characteristic fossils in the Bindi limestone are Spirifers—such does not appear to be the case in the Tambo series, nor is Lepidodendron apparently known there, the Avon River more to the S.E. being the chief habitat of that plant.

Mr. Howitt has, I think, clearly shown that the Bindi beds are below the Tambo. As to the Porphyries, they seem in places to belong to the Granites (which at Moamba, in N.S.W., are stanniferous), and occupy a very prominent feature on the long

---

Peel River, in New South Wales, it occurs in fine gray sandstone, with Ferns and Sigillaria in close proximity to beds of Marine fossils which are as old as Lower Carboniferous. It occurs also about 10 miles N.W. of Goulburn, and Devonian Marine fossils are known to exist not very far off in the County of Argyle. It has been reported also on the Warrego River.

Besides these fossiliferous evidences of supposed Devonian age, there are beds of grit, sandstone, and conglomerate occupying positions of extreme doubtfulness as to age, not only in Victoria but also all along the coast ranges of New South Wales, which, as described by me and confirmed by Mr. Daintree, are certainly older than some parts of the Carboniferous formation. They make a near approach to the "Old Red" of Europe. In my "*Report to the Government of New South Wales*" (6th March, 1852), I have mentioned that I had traced these beds "from the head of the Shoalhaven to the head of the Genoa"; and Mr. Daintree, in his Report to Mr. Selwyn, Director of the Victorian Survey (26th May, 1863), adopts my description, word for word, as applicable to "the Grampian sandstones, the conglomerates south of Mount Macedon, of the Avon River and Tambo, Gippsland"; and he adds, "there can be little doubt they are all members of one great formation."

At Mount Tambo, according to Mr. Selwyn (1866), they underlie the limestone of that locality, which he therefore considers as probably Carboniferous; and this, as stated above, was my view in 1851. (From 2nd. Ed., pp. 8–9.)

descent of Jacob's Pass to the Tongaro River. In 1851 I considered the Granites and Porphyry to be Devonian, and I know now from my own researches and the revelations of Mr. Howitt and others that bedded Devonian rocks may be traced at intervals in a somewhat direct course from Gippsland to the County of Phillip.

There are several important deductions in Mr. Howitt's paper which there is no space here to consider. It will be of great value to any one interested in the study of the Palæozoic formations of Australia, especially the relations, supposed or real, between the Middle and Upper divisions of them.

I cannot refrain from noticing here the service rendered to this question by my friend C. S. Wilkinson, F.G.S., who has lately brought out a map, under the auspices of the Department of Mines at Sydney, of a tract of country intimately known to myself during the last thirty-seven years, and previously alluded to (p. 17), showing the geology of Hartley, Bowenfells, Wallerawang, and Rydal, and the relations of the Upper and Lower Carboniferous, Devonian, and in part Upper Silurian formations, together with Granite, &c., in that part of the County of Cook which surrounds the Western Railway from Hartley Vale to the County of Roxburgh. It was in this area that I first found gold in February, 1841, and in which (and recently in Mr. Wilkinson's company) I have renewed my researches in geology from time to time.

As it belongs to the topic immediately in hand, I consider it only a duty (after so long an acquaintance with the country delineated) to testify to the general accuracy of the details, and the carefulness with which they have been expressed. It is the first work of the kind which has emanated from this Colony, and is at once a proof of the skill and honesty of the author, and a credit to the country. This map alone will serve to refute the absurd statement made in "*Progress Report of the Geological Survey of Victoria*," No. III, 1876, p. 62, that "*Devonian rocks have not been discovered elsewhere in Australia*" than in Victoria! without referring to De Koninck's account of the numerous fossils of that age collected before 1850 in various parts of New South Wales, and which in the very year (1876) when the dogma was proclaimed *ex cathedrâ* had been described, figured, read before an eminent Society in Europe, and proclaimed by publication to the world!! (*See Appendix XV.*)

§ 4. UPPER PALÆOZOIC.

Notwithstanding the opinion expressed respecting *L. nothum*, I do not however affirm that Lepidodendroid plants do not occur in our Lower Coal Measures, as I have for years affirmed it;

and even *L. nothum* may, for anything I know to the contrary, ascend to them and belong to both Upper and Middle Palæozoic. In the section on the Devonians facts are mentioned which show that such plants are well known in our Carboniferous beds, and there are numerous others which can be easily established.

Other acknowledged Lower Carboniferous plants are also known, though denied in no very gracious spirit some years since. Professor M'Coy, as we have seen, doubted it, and De Zigno accepted the doubt; but the plants are here nevertheless, and were not manufactured out of Mesozoic specimens.

Mr. Leo Lesquereux, of Columbus, Ohio, whose reputation is sufficient authority, was good enough to examine two carefully photographed examples from the Rouchel River which I sent to Professor Dana, and pronounced one to be *L. dichotomum* and the other *L. rimosum* of Sternberg.

Professor De Koninck also found embedded with the Marine fossils of my collection from the Lower Carboniferous of Muree, Glen William, Burragood, and the Karua, various well-known plants, such as *L. veltheimianum* (Sternb.); Bornia *radiata* (A. Brongn.); Calamites *varians* (Germar); Schizopteris (sp.), &c., &c.

Dr. Feistmantel (Palæontologist to the Geological Survey of India) has also recognized in the strata from Smith's Creek, Stroud, Rhacopteris (first found there by me in 1850); Tæniopteris, (near) *Eckardi* (Germar); Cyclostigma *Kiltorkanum*; a *Palæozoic* Sphenophyllum, with *Glossopteris*, which also occurs in some of the other Lower strata, as at Muree, &c., Lepidodendron, &c. ("*Records of G. S. of India*," No. 4, 1876.) I privately learn that Dr. F. thinks the species of Rhacopteris are very near to R. *transitionis* (Stur) and R. *flabellifera* (Stur) [= Cyclopteris *inæquilatera* (Göpp)] and near to Sphenopt. *Römeri* from Rothwaltersdorf in Silesia—*Culm.*

I have now sent additional specimens to Calcutta, and the question of *Otopteris* from Arowa will be settled which I have written with (?) as reported from Stroud, it being quite possible that Rhacopteris may occur in each locality. (See *Appendix X.*)

One object in quoting these data will be served by comparing them with the extracts in the note below. *

---

* The Baron de Zigno having stated in 1860 that the Indian strata with fossil plants belonged to the "Lower group of the Oolite," adds :—"This would not be the case with those of Australia, if the observations made in 1847 by the Rev. Mr. Clarke were confirmed, for he mentions in these deposits the presence of the genera Sigillaria, Lepidodendron, and Stigmaria, which would settle the question. But I am not aware that the facts thus cited have been since verified. On the contrary, no mention is made of these genera in the works of Messrs. Moore and M'Coy, in which we are presented with a series of forms which, together with local types analogous to those of India, there are species which recall the Jurassic flora of Scarborough." [" *Some Observations on the Flora of the Oolite. By Baron A. de Zigno,*" Q.J.G.S.

Having personally compared with specimens from Kiltorkan (in my possession) the *Syringodendron dichotomum* (of Mr. Carruthers's paper before referred to) which I sent home to England some years since, and which is yet in the Geological Society's Museum, let me add that I found it in company with the *Lepidodendron nothum* and some other casts of plants in the year 1852.

I would remark, that in one locality in Tasmania I collected many individuals of a species of so called Syringodendron, which occurred in the Coal Measures at the base of Spring Hill, on the slope of which hill Strzelecki stated that he found in beds of sandstone *Pecopteris odontopteröides* underlying the *Pachydomus globosus*, known to Professor M'Coy as a Wollongong Lower Carboniferous shell. It is only fair to add, that though I made in two different years a close examination of the hill and the surrounding district, I failed to recognize the shell, though I saw much that reminded me of the geology of certain parts of the Hunter River Coal formation, and of the Illawarra, of the age of which there is little doubt. I have lately learned that at least two Marine fossils above the plants have been found on Spring Hill.

On the borders of the Devonian formation in parts of the Hunter and Manning River basins, the Lower Carboniferous which is highly inclined passes on along the same strike into beds charged with *Lepidodendron, Knorria, Sigillaria*, &c., and in some instances Lepidodendron occurs in the same blocks with ?*Otopteris ovata* of M'Coy, an example of which was shown in the Exhibition at Sydney in April, 1875, from the east of Stroud. On the ranges at the head of the Peel, and about Booral, Stroud, and Scone, occur numerous fragmentary blocks with Lepidodendron, Sigillaria, and other usually associated fossils of Carboniferous beds.

These and other facts of similar kind have been often stated by me on former occasions. They are referred to on this, in order to show the relations of the New South Wales formations. At present many of the points where the Upper and Middle Palæozoics meet are ill-defined, and it will require the researches and labours of many years to fill them in with strict accuracy.

---

xvi, p. 111.] In reply to this, and to some very remarkable attempts by another critic to show that I had made out my Palæozoic species from fragments of Mesozoic plants (on which I do not choose to comment), I confine myself to one further extract from a paper written by me in 1860, and published in March, 1861 :—" I placed years ago in the Australian Museum at Sydney, specimens of these disputed plants, and in the present year I saw one of the species in the University Museum at Melbourne, which had been found in Gippsland. [" *On the relative Positions of certain Plants in the Coal-bearing Beds of Australia:* By Rev. W. B. Clarke, *M.A., F.G.S.*"; Q.J.G.S. lxvii., p. 355.]

Nor can it be wondered at, that in so large a territory, and with such complicated and broken features, details must for a long period to come give way to generalizations. That the two formations seem to have a passage from one to the other is pointed out by numerous instances in this Colony; and it may be illustrated by the occurrence of Bornia *radiata* (see p. 30), which is one of the distinguishing fossils of Schimper's "*Epoque Paléanthracitique*" intermediate between the two (see *Tome* iii p. 620), as well as by Rhacopteris which is so common at Smith's Creek, Stroud; nevertheless, M. De Koninck considers the mass of the Lower Carboniferous Marine beds he has described to represent only the Upper and Middle Carboniferous of Europe. Other instances of like kind could be pointed out of passages from the one formation to the other. And this I have endeavoured to establish in relation to connection of our Coal-seams on the Hunter and elsewhere with a Palæozoic series through the occurrence of genera of plants which have generally speaking a Mesozoic character.

One aim of my labours in Australia has been to show that we have a succession of groups passing upwards so as to present collectively one great series of Coal-bearing beds, instead of an interrupted widely separated series of formations which have no connection with each other. The question between Professor M'Coy and myself was precisely of this character. He has held that, "no real connection" exists between the beds under the Coal and the Coal Measures themselves; but that " they belong to widely different geological systems, the former one referable to the base of the Carboniferous system, the latter to the Oolitic, and neither showing the slightest tendency to a confusion of type."

It is quite true that there may be succession of formations one over another without any ascertainable break; and there may be also in any given series of deposits of one age interruptions and partial dislocations, marking time in the deposit of the strata without any actual change of epoch; and in cases there may be a want of parallelism or "conformity" in the beds of one and the same group; and it is conceivable how plants that have grown in one age in one country may be found to have grown at another age in a distant territory, without the means of our tracing the missing links in their connection. And, therefore, such plants as Pecopteris *Williamsoni* might be found in China, or in Australia, as well as in Yorkshire, with or without any inference as to oneness of epoch; and very much discrimination and labour may be necessary to discover that the plants are identical and have collateral evidence, derived from the correspondence of matrices, clays, sands, or shales in which they are imbedded, that they grew near where they are found, under the same actual climatal or physiological conditions.

It is this required proof which renders it, in some cases, no wilful scepticism to call in question the identification of any plants; and it may be questioned whether any of our fossils are capable of such complete identification as to make their recognition a matter of positive certainty. On the other hand, even in organic remains of Marine origin there may be difficulties of another kind. We need only refer to the very interesting report of the meeting of Members of the Geological Society of France at Roanne in 1873,* to see how a plant, noticed above (*Bornia radiata*), occurs in the Coal-beds near Roanne, in a variety of deposits all apparently of one formation, which M. Douvillé shows to comprehend in itself two distinct and independent formations, exhibiting singular and extraordinary concordance and discordance, owing to certain physical derangements which bring as it were the strata of Sarrebruck Coal-field and that of Saxony into Central France, and, at the same time, according to M. De Rouville, exhibit a concordance between the Coal formation and the Permian, and discordance between the mountain limestone and the true Coal-beds without mixture of the flora. After well considering such a condition of things as this, one should be very cautious in the matter of stratification, especially in a country so distant from Europe and America as is Australia. It was my lot to pass through the district in question in 1825, and I still retain impressions of the geology, so far as I noticed it; but I regret that I did not know it, as I now find, to be in many more points than those just mentioned like that of some portions of our Australian Coal-fields. A more striking instance of the conformity of stratification of widely separated formations may be found in a memoir of Casiano de Prado, "*Sur l'existence de la Faune Primordiale dans la chaine Cantabrique*," read by M. de Verneuil before the same Society, on 7th May, 1860,—in which is shown a section of vertical rocks perfectly conformable to each other without the slightest break, Lower Devonian rock and Red Sandstone with the Devonian fossils side by side and on both sides of a band of Lower Silurian, as determined by Barrande and De Verneuil, whose descriptions of the fossils is given in the memoir. The whole of the rocks mentioned, which are succeeded by carboniferous strata equally vertical, were considered at one time to be Devonian. On this most instructive example M. Barrande makes the following excellent remark, which will be sufficient apology on my part for calling attention to the necessity of carefully examining the Stratigraphy as well as the Palæontology of the rocks in Australia,—"This example is so important in its results that it deserves to be cited in the number of those which

---

* "*Bulletin*," 3ᵉ serie, tome 1, pp. 441–450.

prove how the mutual aids ought to wait on each other, of the two principal branches of Geology, that is to say, Stratigraphy and Palæontology. Their comparative application could only be despised by minds prejudiced and disposed to sacrifice the progress of science to the ephemeral maintenance of their exclusive and systematic views,"—the conclusion of the paragraph I leave in the original, "*Des esprits si étroits ne se trouvent pas parmi nous.*" ["*Bull. Soc. Géol. de France,*" 2ᵉ ser. xvii, p. 543.]

In the course of my work I have had to contend with the prejudices of some who *have never visited this territory*, and who, from a distance of many hundred miles, have ventured to dogmatise, solely from a palæontological point of view, without caring to ascertain how far the stratigraphical evidence is at variance with their conclusions.

In consequence of this the ascending order of formations above the Lower Carboniferous in this Colony has long been disputed by some, whose unacquaintance with facts patent to all who have examined them is the best apology for a more temperate style of criticism than has been adopted.

We are, however, indebted to Professor M'Coy, for ascertaining, in 1847, the existence of eighty-three species of animal remains in our Carboniferous formation, in a collection forwarded by me to the University of Cambridge, in which the Professor was then officially employed.

Before that time, Bowerbank, Sowerby, Morris, and Dana had determined the existence of the Carboniferous Marine beds; and the latter author enumerates about eighty species observed during his excursions in New South Wales, in some of which I accompanied him. (*See Appendix II.*)

More recently Mr. Etheridge has described fifteen species of Lower Carboniferous fossils from Queensland, in relation to Mr. Daintree's paper on the geology of that Colony, of which ten were furnished by myself. None have yet been discovered in Victoria. In Tasmania, Mr. Gould figured some well-known forms from that Colony, but the plates were never published.

He has noticed also, what I have contended for, that the worked Coal-beds of the Mersey River belong to the same formation with Palæozoic Marine fossils, as in Queensland and on the Hunter River.

Having visited the Tasmanian locality for the purpose of inspection, I can confirm all that has been stated respecting the occurrence of the Palæozoic fossils, Orthonota, Spirifera, Fenestella, Pachydomus, Theca, &c., in association with and immediately above the Coal; and lately I have been officially informed that Coal-seams have been found by piercing these beds on the Don River, confirming my grounds for recommendation to look for them.

In Western Australia traces of these Marine beds have been detected and announced by Mr. Gregory. And in extension of the formation northwards beyond the limits of Australia, it is well known by more than one observer that Carboniferous beds exist in the Island of Timor, where Beyrick discovered several of our New South Wales species, *e.g.*, *Spirifer lineatus, Sp., Tasmaniensis, Productus semireticulatus, P. punctatus, &c.* ("*Acad. des Sciences de Berlin,*" 1861.)

My own collections received in 1874 from Queensland some interesting additions, which arrived too late to form part of the contribution to the Daintree Collection.

Among these fossils, from the head of Bee Creek (Fort Cooper), I find Pecten, Spirifer, Trochus, and magnificent specimens of Productus, and a variety of usually associated shells, and with them in the same brown ferruginous grit and shale-beds well-depicted *Glossopteris*, and some other plants, one fragment of which appears to be of a *Dictyopteris*.

This mention of Glossopteris will lead to considerable discussion respecting its occurrence in beds interpolating in the Marine fossiliferous strata, as well as occurring in the shales of the Coal-seams on the Hunter River and elsewhere in New South Wales and in Queensland.

Mr. Daintree, F.G.S., ("*Notes on Geology of Queensland,*" Q.J.G.S. xxviii, p. 286) gives a section from the Coal-seams near the Nebo Crossing of the Bowen River, of Coal-seams with fragments of Glossopteris underlying Productus and Spirifer beds several hundred feet thick, with abundance of other Carboniferous mollusca, the strata being upheaved by porphyry, and the lower beds resting upon it. The beds are "quite conformable." At the junction of the creek the argillaceous Marine beds are surmounted by others consisting of "coarse grits and sandstones with interstratified shales. In these the impressions of *Glossopteris* are very common and sometimes beautifully preserved; but," he adds, " I have never been able to find the fauna and flora unmistakeably represented in the same bed."

This admission as to that particular locality has been used against the evidence of the beds themselves as to their position by the author of "*Report of Progress by R. B. Smyth, No. III,*" 1876, p. 59, who states that "any misapprehension in regard to the age of the Mesozoic Coals of New South Wales, probably, is due to the accidental or apparent conformableness of the Mesozoic strata to the underlying Carboniferous (Palæozoic) rocks" !! In addition to the evidence from the junction of Nebo Creek and Bowen River, Mr. Daintree cites, with a section, the facts observed at Pelican Creek, where he shows "Marine beds resting directly on a Coal seam." "At the base of this cliff a seam of Coal about 4 feet thick crops out the entire length of the section.

c

Directly upon this rests a coarse-grained sandstone, with a few imperfect casts of shells; while at the top of the cliff an arenaceous limestone band holds abundant specimens of the *Streptorhyncus crenistria* so common throughout all the Lower Marine series." (p. 288.)

Here we have in addition to Glossopteris below Marine beds, Coal also below them, and lower down Glossopteris in the sandstone and shale—facts quite in keeping with what has been so clearly shown by myself in parts of the Hunter River basin in New South Wales.

In further confirmation of such evidence, I will now quote an extract from a letter by Mr. Daintree, dated "Maryvale, North Kennedy (Queensland), January 22nd, 1870. On the M'Kenzie River, near the junction of the Isaacs, the Coal Measures are highly inclined, *Glossopteris* the common fossil; but running up Roper's Creek they gradually become horizontal, and at the top of the Roper's Creek watershed horizontal beds of sandstone and sandy limestone are the only rocks exposed in section full of Hunter River fossils, *Producti*, &c. * * * I could only be more assured than ever that the *Glossopteris beds underlie these horizontal Productus beds*, and a week spent in surveys would altogether settle the matter. I see Hector gets Glossopteris associated with Mesozoic fauna in New Zealand; I am satisfied we have it with *Palæozoic Carboniferous* fauna."

Speaking of the valley of the Comet Creek, Leichhardt ("*Overland Expedition*," pp. 104–5) says he met, on January 9–10, 1845, with sandstones in the deep gullies running to the creek and on slight elevations—"Sandstone crops out in the gullies of the valley in horizontal strata, some of which are hard and good for building, others like the blue-clay beds of Newcastle, with the impression of fern leaves identical with those of that formation. At the junction of Comet Creek and the river I found waterworn fragments of good Coal and large trunks of trees changed into ironstone. I called this river the 'M'Kenzie,'" in honour of Evan M'Kenzie.

It is so far certain that the Newcastle beds underlie the Marine Carboniferous near the junction of the Comet and M'Kenzie. Gregory found the remains of Leichhardt's camp on 17th November, 1856, but records no geological data at that spot (J.R.G.S., xxviii, p. 128). But W. Lockhart Morton ("*Notes on Northern District of Queensland*," Trans. Phil. Inst. Victoria, read 23rd January, 1860) noticed the Coal in large angular blocks at the junction. In another spot some distance from the junction and on the M'Kenzie he observed "in a stratum of sandstone an angular piece of beautiful bright Coal embedded—proving that this piece of Coal is of greater age than the sandstone and than the seams of Coal which that sandstone overlies." (p. 13.)

Respecting Mr. Daintree's evidence, may be added here the extracts from two letters from that gentleman to myself (which were published by me in a paper on "*Marine Fossiliferous Secondary Formations in Australia*" (Q.J.G.S., xxiii, p. 11) :—

"Bowen, February 10, 1866.
"In the Bowen River Coal-field, your statement as to the Palæozoic age of the Newcastle beds is, so far as I could judge, *entirely proven*, since we have *Spirifers*, &c., similar to those in Russell's shaft and the railway section at Maitland overlying the Coal-seams, *Glossopteris* being the most abundant fossil fern."

"Brisbane, April 11, 1866.
"I send you a copy of what Professor M'Coy addressed to me after an examination of the fossils I took him, viz.,—
'Your brown beds No. 2 are identical with the Marine beds *underlying the Coal of the Hunter*' [*i.e. overlying* the Stony Creek Coal-seams.—W.B.C.], 'the Productus *brachythærus, &c., &c.*, fixing them. The Streptorhynchus is new, but of clearly carboniferous type. I have no doubt of their being Upper Palæozoic.
'The plants are *Phyllotheca Australis* and *Glossopteris Browniana*, forms related to which in Europe are only found in Mesozoic rocks.'"

As to Lepidodendron, I have no where asserted that the Lepidodendron, Sigillaria and other plants of that class have been found in or over the beds of Newcastle or at Wollongong, though I have mentioned already the possible discovery hereafter; but I have asserted many times that such plants occur in some parts of our Coal Measures, and that below the Marine fossils which underlie the Upper Measures *Glossopteris* occurs, and others which have been by some considered solely of Mesozoic age; and I have therefore argued that there *is* "a connection," which has been denied, the denial in my opinion having arisen from want of personal experience on the part of my opponents, though I have given them the same credit I take to myself, viz., that we each and all come to conclusions to which we are led by our individual acquaintance with or ignorance of facts. That some of the doubters have contented themselves with passing sentence without sufficient inquiry is distinctly stated by one of them in a Parliamentary document, from which extracts will be found further on. I refer now to the "*Progress Report from the Select Committee on Coal Fields*," Melbourne, ordered to be printed, 20th October, 1857, and to the Evidence under questions Nos. 461, 471, 577, 581, 582, 584, 586, 588 to 592. No one who peruses that evidence will deny that it was upon preconceived Palæontological determinations *alone*, without the condescension of a local research, that positive dogmatic dicta were declared as law with a wilful resolve to over-rule any opinion in opposition. Mr.

Carruthers, whose judgment none perhaps would rashly call into question, in the discussion which ensued upon the reading of Mr. Daintree's paper (op. cit.), argued that "With regard to the supposed Glossopteris and Tæniopteris Epochs, which by some had been regarded the one as Palæozoic and the other as Mesozoic, he was not convinced that they could be distinctly separated, but thought rather that they might belong to *different portions of one great period.* * * * * He thought that neither was of a date earlier than Permian."

The conclusion I have all along held is, that the "Carboniferous" strata, and those which it pleases dissidents to fancifully designate as "Carbonaceous" (which is, at the best, a *misnomer*), are parts of one great series, and that the beds which contain Mesozoic Marine fossils may be properly placed in still higher stages of the Palæontological edifice.

In noticing my opinion expressed in 1861 (in a paper "*On Recent Geological Discoveries and Correlation of Australasian Formations with those of Europe*"), Sir Roderick Murchison ("*Address to the Geol. Sec. of Brit. Association at Manchester,*" Sept. 5th, 1861), holding the view of a *possible* double series, stated that he had received a communication from Mr. Gould in which he (Mr. G.) says, that in "Coal-fields of the rivers Mersey and Don, one of the very few which are worked in Tasmania, he has convinced himself that the Coal underlies beds containing specimens of true Old Carboniferous fossils," and adds "that in Tasmania at least, the Coal most worked is unquestionably of Palæozoic age," (p. 23.) In this Mr. Selwyn (Q.J.G.S., xvi, p. 147) fully concurs.

Now, in the paper on which the above comments were made I had expressly affirmed, that reviewing the whole discussion I was willing to admit, "that though some of our Coal appears to belong to the true Carboniferous epoch, yet it is possible that some may belong to the Permian epoch as suggested by Mr. Dana for the Newcastle Coal, or to the Triassic as suggested for the Indian and Virginian Coal; but I am not yet [*i. e.*, in 1861, *nor am I now*, 1878] convinced that our New South Wales Coal-seams are of *Oolitic* age."

My highly respected friend Dana at one time abandoned the Permian for the Trias, and Dr. O. Feistmantel of Calcutta is labouring diligently to support this view in opposition to those of Dr. Oldham and Professor Blanford. (*See p.* 62.)

But the question is still an open one, although the Oolites are insolvent; and if all our N. S. Wales Coal is not somewhere between the Trias and Palæozoic, or at the top of the latter, in an intermediate Palæozoic stage not known in Europe, it will require strong faith and stronger affirmation to cast it *all* into a Mesozoic receptacle, notwithstanding the possible reliance of my Victorian friends and critics on one or both of the assertions that there is no connection " between the Newcastle Coal and the *base* of the Carboniferous" (which may be true as far as the *base* is concerned), or that it is " not older than the Trias, nor younger than the Oolite"—so that if the Trias wins, Oolite is no where! Q.E.D.

Those who deny the asserted age of our workable Coal-seams affect to rely on the assumed age of that most prominent plant, *Glossopteris Browniana.* They say Glossopteris is an Oolitic genus, " *Exactly as in the English beds the Glossopteris is associated with Tæniopteris?*" i.e., in the assumed Oolitic series. To this we may reply, that " *Glossopteris Browniana*" which is " *the* Glossopteris" alluded to in the above extract from the " *Report of the Three Commissioners on the Western Port Coal-fields,*" (p. 8), is a plant *utterly unknown* in Europe and America, and only known in India, South Africa, and Australia; and that Tæniopteris, which is said to be associated with it in English beds, according to Schimper, the most recent expounder of Fossil Botany, is a genus which has only five species, all of which are Permian, *i.e.*, of Palæozoic age or of Upper Carboniferous. Even if one Tæniopteris should be found in the same beds with Glossopteris, that fact would not invalidate but would rather strengthen my argument. Since the former is Palæozoic, and the latter occurs in the Coal-seams below the beds which are filled with Lower Carboniferous Marine fossils, it is clear that those beds and the plant they hold must certainly be Palæozoic, whatever becomes of any other part in the succession of the series or group to which they belong. It was attempted to be shown that there exists an inversion of beds at Stony Creek, where five seams of Coal holding Glossopteris, under 143 feet of acknowledged Palæozoic Marine beds, occur (the fossils from which I sent down to Sir Henry Barkly, who submitted them to Prof. M'Coy), and to meet this I requested that a geologist might be sent up from Victoria to test the facts. Accordingly Mr. Daintree came, and in the " *Yeoman,*" a Melbourne journal, No. 100, will be found his refutation of the inversion story and a full confirmation of my assertion. This circumstance is ignored by the Commissioners of 1872, as are all others that do not fall in with the imagination of certain critics in Victoria. But I may now add that Glossopteris in Coal-seams below the Marine beds has been found in other localities, as for instance at Greta, where the Coal has been reached below more than 400 feet of Marine strata, Glossopteris and other plants also occurring 2 feet 6 inches above the Coal. (See *Sections No.* 1 *and No.* 2 *at the end of this Memoir.*)

Not only so, but it is found in sandstones elsewhere, amidst the Marine fossils themselves, and in the very same portions of rock with the latter. So that no reasonable doubt ought to exist in the mind of an honest controversialist that "*Glossopteris*" does occur as early as the so-called Lower Carboniferous strata, and therefore our Coal-seams have a right to be held of that age.

Now Schimper, to whom I before alluded, considers that the Indian, African, and Australian plants are merely varieties of the same *G. Browniana*. In India no Marine fossils have yet been found in connection with its Coal plants; and in Africa the Glossopteris is not set down to any older formation than Triassic by Mr. Tate; but even that is older (although Mesozoic) than Oolitic, to the latter of which M'Coy refers them. And if Glossopteris has a range as extensive as some other fossils which pass through three separate series of strata, why might not it pass up into Secondary rocks, without denying its existence in Australian Middle or Lower Carboniferous? *There* it clearly does not govern, but must be subordinate to the *Fauna*. But it is not alone in that position; other plants also occur therein which have as much an Oolitic facies as itself. And yet it is undoubtedly true, as is well shown by Daintree, that in Queensland Glossopteris is confined to beds that are in association with Palæozoic fauna, and that the so called Tæniopteris is found to accompany a Mesozoic fauna; and I can aver, after thirty-nine years experience, that no Marine deposits of Secondary age have yet been discovered in New South Wales, although in Queensland beds of Coal occur in supposed connection with such.

There may, therefore, be two epochs of Coal as suggested by Murchison or as stated by Mr. Carruthers two portions of one series, without dispossessing the lower portion of its right to hold a property in a plant that may not have existed in the time of the younger part of the series. Whatever be the value or uselessness of reasoning on the point, this fact still remains— *Glossopteris Browniana* does exist in New South Wales and in Queensland in Coal Measures that interpolate strata full of Palæozoic Marine fossils; and is absent in the latter Colony, where the Marine accompaniments are called Mesozoic, and does not exist at all, so far as is yet known, in Victoria, where the Palæozoic and other Marine beds are at present missing.

As to the division arbitrarily made by Professor M'Coy, in a list re-arranged by him of Mr. Keene's specimens, separating "*Shale with G. Browniana and Otopteris*" from the Palæozoic beds, that excellent Palæontologist may be assured that a plant apparently the same as Otopteris (? *ovata*) is combined with Lepidodendron, Rhacopteris, and other plants near Stroud; and that at Greta, and at Mount Wingen, Glossopteris is found below his own determined Palæozoic Marine fossils, the smoke from the burning seams full of the plant at the latter locality passing up through cracks in the overlying conglomerate full of Palæozoic shells, &c.

Nor does the arrangement made of Mr. Keene's collection agree with the actual facts in nature, for the Greta beds are not the uppermost with Marine fossils, but beds with them lie further to the east—in which Phyllotheca has occurred at Harpur's Hill, and Glossopteris in the same way at Muree near Raymond Terrace.

There is another item to be taken into account—the occurrence of fishes—one in the Newcastle seams, described and figured by Dana, viz., *Urosthenes Australis;* and many of different species in the beds above the Coal Measures, of which mention will be made hereafter.

The greater part of them are fragmentary, but others are entire. Some specimens exhibit the head, others the tail or hinder portion of the body, and one jaw has been found with teeth which are not shown in the other fishes. One has within a few weeks (January, 1878) been found in shale at Balmain, near Sydney, which is shaped like Belonostomus; but the scales are not shown and the caudal fin is too indistinct to be traced, the vertebral column and some of the ribs are better defined, but it is out of shape, and can be merely guessed at. Only half of the body was found in digging a well, and the remainder was searched for, at the desire of Mr. Wilkinson, F.G.S., and myself, and has been with difficulty discovered.

Of these fishes, Palæoniscus was recognized by Sir P. Egerton, from Parramatta and the Gibraltar Tunnel, between Nattai and Bowral, and one species assigned by that learned Icthyologist is *P. antipodeus.* He considers the fishes to be Permian (Q.J.G.S., xx). Professor M'Coy also has admitted that they have "a general aspect of Triassic or Permian fishes." ("*Official Record,* 1866-7, *Melbourne Exhib.,* p. 169.")

Some recent writers have called in question the claims of Palæoniscus to any Carboniferous rank. In a learned paper by Traquair—(Q.J.G.S, xxxiii, pp. 548-578. "*On the Agassizian genera Amblypterus, Palæoniscus, Gyrolepis, and Pygopterus. By Ramsay H. Traquair, M.D., F.R.S.E., F.G.S., Keeper of the Natural-History Collection in the Edinburgh Museum of Science and Art.*" Read May, 1877)—the author, after great detail and illustration, comes to the conclusion that the fishes named are not Carboniferous but Permian.

This determination does not interfere with the view I have maintained of the Palæozoic age of some at least of our Australian Coal Measures, which *is* half supported by M'Coy, wholly by Egerton, and again confirmed by Dana. Mr. W. J. Barkas, L.R.C.P.L. and M.R.C.S.E., in a paper read before the Royal Society of New South Wales, 3 Dec., 1877—("*On a Dental peculiarity of the Lepidosteidæ*")—devoted himself to a consideration of the doubtful character, as he considers it, of the fishes, by Sir P. Egerton, viz., Palæoniscus, Urosthenes, and Myriolepis.

Mr. Barkas says there is no description of Urosthenes. This is, however, given by Prof. J. D. Dana, in the "*Geology of the U. S. Exploring Expedition* 1838-1842, *under the Command of Captain Wilkes, U.S.N.*" (New York), p. 681. His chief ground of inquiry is respecting the teeth of these fishes, as he considers the Lepidosteidæ are distinguished by the teeth being "tipped;" and, unfortunately, as in many instances in Europe, the New South Wales fishes alluded to show no teeth whatever. But, as stated in the last edition of this memoir, p. 39, "the last specimen of fish from the Palæoniscus beds, reported by me to Sir Philip Egerton, was a portion of a jaw of a fish whose teeth were of a Saurichthyan type, but the learned Icthyologist considered it also to be Permian." The teeth in this specimen were so completely "tipped" in the way mentioned, that I considered it to be a *Saurichthys* (see Agassiz, "*Poissons Fossiles*," vol. ii, tab. 55a.), and named it as such, under correction, to Sir Philip.

The objection of Mr. Barkas may therefore be considered as answered; although the fish to which the jaw belonged is not precisely known, Mr. B. says, "from the writings of Professors Owen and Agassiz, I learn that Saurichthys is also tipped with enamel," (op. cit., p. 205). As he admits, moreover, that of the eighteen genera Lepidostean which he cites, ten are "tipped," it is probable that the Ganoid fishes discovered by me, were also "tipped," though no teeth have been found. I was not present at the reading of the paper by Mr. Barkas, and did not read it in print till 9th February, 1878, or I would have replied at the time.

The peculiarity of the teeth in Permian Ganoids was long ago pointed out by Dr. King (See "*Monograph of the Permian Fossils of England:*" *Palæont. Soc.*, 1850, p. 228, under *Platysomus macrurus*, p. xxvi, 1.)

Then, as to the "*vulgar error*" that heterocercal ganoid fishes are confined to Palæozoic beds,—which any one acquainted with ordinary treatises on the subject may be supposed to understand is an error, though scarcely "*vulgar*" in the ordinary sense of that often offensively used term,—surely it may be permitted to conclude from the fact that among all the fishes discovered in our Coal-beds, and in the beds above the Coal, not a single homocercal tail has been found, the probability is, as Sir P. Egerton has surmised after examination of those submitted to him, that the *fishes are Palæozoic*, especially as the admission is made that "*the homocercal* structure is *not* known in Palæozoic rocks." ("*Report on Coal Fields.*" Victoria, 1872, p. 6.)

The fact that the Coal-beds overlie or interpolate the Marine beds in what is called "conformable order," ought to be considered a satisfactory conclusion that no break such as ought to exist under other circumstances does exist, because whether the

Coal Measures are horizontal or inclined they merely follow the same condition in the Upper or Lower Marine beds with which they are always associated.

The argument from the occurrence of fish remains is met by the incidental remark that the "heterocercal ganoid fishes being of genera and species *peculiar to the locality* have no value" in determining the age of the beds in which they occur, may be met by the retort that if *peculiarity* is to be a guide in determining geological age, there is an end of any certainty for such persons as affect to uphold their own theories by reference to peculiar plants; and this Professor M'Coy himself does in relation to a Scarborough plant by which he affects to guide his Oolitic determination to the exclusion of Glossopteris and its usual associates.

Respecting Palæoniscus, one of the New South Wales fishes, a passage translated from Agassiz, whose decision ought to be satisfactory, will not be out of place, considering that it meets the objection on the form of the caudal fin. He says,—" I know ten species of this genus, which appear to be limited to Coal Measures and the Zechstein. It might not, however, be impossible to discover traces in the *Grès bigarré*,* the Muschelkalk, and the Keuper" (*i.e.*, in the Trias); "but that which I believe I am able to affirm is, that *it does not ascend to the Jurassic formations*, of which the numerous representatives of the order of ganoids have the tail *regular*, and never prolonged in a long point forming the upper lobe of the caudal, as takes place constantly in the genera of the earlier formations. I do not understand what were the intentions of Nature which have produced these singular differences, but it is certain that they exist, and it would be to misunderstand our duty to ignore them, or to attribute less importance to so general and constant a fact." ("*Recherches sur les Poissons Fossiles*," tom. 1, p. 43.) To this may be added, that the generality of the fishes in N. S. Wales are heterocercal;

---

* He afterwards names *P. catopterus* as belonging to this sandstone. It was, however, only found in one spot, only "a few square feet" in extent, in the county of Tyrone: (*Portlock*: "*Geology of Londonderry, Tyrone, and Fermanagh*, p. 468.") Respecting this fish, Dr. Traquair says (Q.J.G S. xxxiii, p. 565, op. cit.)—"This little species from the Triassic red sandstone of Rhone Hill, county of Tyrone, was originally named by Agassiz, but was not described by him. Sir C. Lyell, however, in referring to it in connection with certain American Triassic forms, says, concerning it,—'The Irish Palæoniscus catopterus of Roan or Rhone Hill, referred by Col. Portlock to the Trias, is a true Palæoniscus, and not allied generically either to the *Ischypterus* of Egerton, or the *Catopterus* of Redfield (Q.J.G.S., iii, p. 278); and in Sir P. Grey-Egerton's brief description of the species (Q.J.G.S., vi., p. 4) occurs the following passage,—'The dorsal fin is placed much nearer the tail than in any other species; in this respect, but in no other, *P. catopterus* resembles the genus *Catopterus* of Mr. Redfield. The tail is decidedly heterocerque'."

though in some instances the caudal fin is not so distinctly pronounced as in others, which may therefore be classed as "semi-heterocerque." But Palæoniscus well developed as to the tail was found in shales and sandstone 1,000 feet geologically above the worked Coal-seams.

The existence of Palæozoic strata of Carboniferous age in some parts of Victoria is, as I believe, a fair assumption of the Cape Paterson Reporters, though at present they cannot prove their position by fossiliferous evidence; but the denial of that existence would hand over their whole Coal-territory to a formation or formations to prove the age of which they have no more marine evidence than they have respecting a Carboniferous era. They have never yet seen a single Marine fossil bed in all Victoria to justify even their adopted view of their Coal belonging to the Oolitic age, which is elsewhere multitudinously fertile in Marine fossils, and this, no doubt, *is* "peculiar." The Reporters on the Western Port Coal-fields notify carefully, that "it should be distinctly understood that *our opinion* respecting the age of the New South Wales Coal Measures is *based entirely* on the collection of rocks, fossils, and Coals forwarded to us by the late Mr. Keene, and on the published reports on these Coal-fields." But even this is accompanied by a sneer at Mr. Keene's alleged blunders in Palæontology.

On the above I would observe that, as I had seen the collection referred to before it was despatched, I am prepared to say it did not completely represent the beds in the local district from which it came, and was only a partial display of the series of the strata in association with Coal throughout the Colony; and that in the arrangement adopted by Professor M'Coy, as quoted in the Report, most important portions of the beds are omitted. I would, therefore, attribute the "opinion of the Board respecting the age of the New South Wales Coal," so authoritatively pronounced, to be based on imperfect data, showing that the gentlemen who then decided the question are *practically* ignorant of the true grounds of decision, clearly not having made any inspection for themselves, and totally ignoring the opinions of the host of observers who have certified to the contrary; amongst whom is Mr. Daintree, a member of the Victorian Geological Survey, the late Mr. Stutchbury, who reported thereon, as well as many others who have studied the strata *in situ*, and are true witnesses against the side of the Oolitical party. In the pleadings on that side, the reliable evidence that makes against them is "burked." and a foregone conclusion is offered as if it were final —and the judgment is delivered *ex cathedrâ*, whilst numerous witnesses of the first credibility are altogether ignored. This may be prudent and ingenious, but it is *not* "*scientific*," nor is it honest, yet it helps to bring out the magnificent declaration:

"We confine ourselves to the *statement* that we have *not before us a particle of evidence* indicating that the Coal-seams now being worked in New South Wales are of Palæozoic age." A great compliment this to persons who have laboured for years to establish truth; but they may console themselves with the reflection, that "*Préjuger est mal juger.*" Amidst this lamentable ingenuity to "tell the truth without telling the *whole* truth and nothing but the truth," and in the arraying of evidence from beyond Australia instead of collecting the whole evidence furnished from itself, there is one grateful exception which, though not entirely satisfactory, is much more so than some previous proceedings were. It would have been better to have acknowledged that old opinions had been re-called.

In the notes on Mr. Keene's specimens, Professor M'Coy, though he draws a line where it ought not to be, has changed his method of putting his old opinions about the Coal itself, inasmuch as he no longer makes use of the notion which he once entertained and put in evidence before a Committee of the Melbourne Parliament. I must explain this: On the 20th November, 1857, he was examined (as the Chairman of a Mining Commission) on the Character and Extent of Coal in Victoria, and he asserted over and over again that no Palæozoic coal existed in Australia. The following answers speak to that point:—

"722. (Answer). The members of the Mining Commission have an impression that, as the Coal deposits to be expected there [Cape Paterson] geologically are not the same as those of the great Coal-fields of England, but are of similar character with the Coal-deposits of New South Wales and Tasmania, therefore *it is unlikely that they will be of commercial value*; and as scientific men they would not on their own responsibility, recommend the expenditure of public money there.

727. (Q.) Considering that the information [? formation] of the Cape Paterson Coal-fields is similar to those of New South Wales and Tasmania, you are of opinion that as an *economic question* you would advise no further prosecution of any surveys in that locality? (A.) That is my opinion.

744. (Q.) You would not advise the prosecution of any further inquiries for the discovery of Coal? (A.) No recommendation to that effect would emanate from myself or the Commission.

747. Such Coal-fields [*i.e.,* those of Palæozoic age], do not exist in this country [*i.e.* in Australia]. That is a point which I wish clearly to show, and I think it is one which has never been clearly shown to this Committee before.

758. I know you are not to expect the old Palæozoic Coal-fields in this part of the world.

759. (Q.) Do you contend that the Mesozoic Coal-fields are not suitable for the different purposes of commerce? (A.) They are not so suitable as the Palæozoic, they are not so extensive, the beds are not so thick or workable, nor is the quality so good over any workable area.

767. (Q.) If a Coal-field at Cape Paterson was discovered equally good with the Sydney Coal-fields, would you consider it worth working? (A.) My individual opinion is that it *would not be worth working.*

771. [Of Cape Paterson] (A.) Of course the Members of the Mining Commission do not wish to attach any scientific weight to their evidence in a commercial point of view, they merely choose to say, that as *men of science,*

no recommendation would emanate from them to undertake extensive works there, *because* the utmost you could expect would be such a *Coal-bed as you have at Sydney.* Once more ; 769 (*By Captain Clarke.*) (Q.) The Virginian Coal-fields of the character you describe as being similar to those here, are worked at 775 feet depth ? (A.) Yes ; but the beds there are *not to be compared to the Palæozoic Coal beds."*

The witness here expressed an undoubted fact, but seems to have forgotten entirely in November, 1857 the evidence he had given before the same Committee on 18th August of same year. *Voici !* By the Chairman :

474. (Question). The Committee desire that no time may be lost, and also to know what aid the Mining Commission can afford them in the prosecution of their inquiries—Are you prepared to offer any facilities for that purpose ? (A.) I have obtained some specimens from surveyors from the Avon Ranges, in the Gippsland district, which is the *first evidence* that the Palæozoic Coal of Europe exists in the Colony. One is a large specimen of Lepidodendron, indicative of this ancient Coal, so that my own opinion is that the principal Coal-deposit to be expected in the Colony would probably extend from the Cape Paterson beds northwards through the Gippsland country, and probably form a union with the Sydney deposits. The Hunter and Hawkesbury deposits of Coal are the finest specimens I have seen of that period. There is reason to expect that deposits of both those geological ages will be found to exist there, so that if arrangements were first made for geological explorations of the Gippsland district valuable results might follow.

Strange to say, however, neither the expectation in 1857 of Coal of the older epoch, nor the denial of its value in favour of that of a " more recent age" after the explorations of a host of skilled surveyors in Victoria, nor the excursive labour of the experienced Examiner of Coal-fields from New South Wales, has yet realised either anticipation in that Colony.* The latest report

* In 1857 the Report from a "Select Committee upon Coal-fields" was ordered, on 2nd October, by the Legislative Assembly of Victoria to be printed. Now, in the evidence given by the witnesses we find the following recorded :—

Alfred R. C. Selwyn, Esq., further examined :—

576. (Q.) *By the Chairman.*—Will you be good enough to read that letter (handing the following paper to the witness) ? " Extracts from Professor M'Coy's letter of the 30th September, 1857, to the Honorable the Chief Secretary. * * * It is desirable to state plainly here the opinion of the Mining Commissioners relative to the expense of trials for Coal, which is, that the richest deposits to be expected in the accessible parts of Victoria would resemble those of Sydney and Tasmania, with this difference, that while the latter are situated most advantageously for the employment of water carriage and cheap labour, the localities in which such deposits may be expected to exist in Victoria are so disadvantageously placed in both these respects, that even if similar rich Coal-beds were to be discovered, the public would not be likely to receive any benefit, as the supply could be more cheaply brought from the neighbouring colonies." (A.) I concur in all that is stated there, except that if numerous thick seams of large extent and good quality were proved to exist, they must be worked to advantage.

577. (Q.) That professes to be an extract from the report from the Mining Commission ? (A.) Yes.

I have seen respecting "Kilcunda and Cape Paterson" is from Mr. Cowan, Mining Surveyor, dated 2nd August, 1875, who, after considerable examination and collection of available information, comes to the conclusion that "very little can be deduced with certainty in regard to either the character or extent of the Kilcunda and Cape Paterson Coal-deposits except by actual experiment. The pick of the miner, will in my opinion, be the only conclusive test." ("*Progress Report No. III,*" 1876, p. 279.)

But the money spent, and the labour bestowed on investigations and search for Coal in Victoria has been enormous, and it is a subject for deep regret that her enterprising Colonists have not been more successful, as a valuable and abundant Coal-field in that Colony would have been, of whatever geological age, most beneficial to thousands of the present and future occupants of that interesting territory.

The old Coal-beds, as well as what the Southern scientists are pleased to call "Carbonaceous" strata, are equally unpromising, and Mr. Howitt shows the reason—because they have been greatly *denuded*.

---

578. (Q.) You are a member of the Mining Commission? (A.) I am.
579. (Q.) Did you sign that report? (A.) No.
580. (Q.) How are meetings of the Mining Commission called? (A.) The Mining Commission consists of Professor M'Coy, Mr. Panton, the Resident Warden at Bendigo, and myself. Mr. Panton is hardly ever in town ; I could not say how many meetings he has attended, but very few ; and no regular meetings have ever been called. Now and then I go up to the University and discuss these matters with Professor M'Coy.
581. (Q.) *By Mr. O'Shanassy.*—In sending in a report from the Mining Commission to the Government, is it the practice to obtain the consent of the other members of the Commission? (A.) Not formally.
582. (Q.) That is, the document is not sent to them? (A.) I have seen the document ; in fact I wrote the report myself with Professor M'Coy, he dictating and I making suggestions, and then it was subsequently copied by a clerk, I suppose under Professor M'Coy's directions, and I have seen it published in the newspapers ; but from the time Professor M'Coy made the rough draft of it I have not seen it ; whether it was ever sent to Mr. Panton I am not aware.
584. (Q.) Does that document meet your views now? (A.) There are some portions of it which do not meet my views.
586. (Q.) *By the Chairman.*—I wish to ascertain precisely as to the constitution of the Mining Commission, you say it consists of three gentlemen, namely, yourself, Professor M'Coy, and Mr. Panton? (A.) Yes.
587. (Q.) Mr. Panton resides at Bendigo? (A.) Yes.
588. (Q.) So that practically you and Professor M'Coy are the Mining Commission? (A.) Yes.
589. (Q.) Is it usual to hold meetings of the Commission? (A.) Not formal or regular meetings of which minutes are kept ; we meet occasionally and discuss things in a manner that I have all along considered was not the way to carry it on.
590. (Q.) Then is it competent for you or for Professor M'Coy to write in the mode you have described a document, and send it in as a report of the

But putting aside all commercial considerations, and returning to the question of epochs, we find the Reporter on the Cape Paterson Coal-fields appealing to China for proof that Coal with Glossopteris and other associated plants in New South Wales cannot be Palæozoic, and in direct contradiction to the opinion of the Palæontologist of Victoria, as stated in the reply, No. 759, (quoted in p. 43), that Mesozoic Coal is not be compared with Palæozoic, treating somewhat neglectfully the value assigned to the Cape Paterson Coal by the Board.

In the Report on the Coal-fields of Western Port, 1872, there are quotations from a letter of Dr. Newberry to Professor R. Pumpelly, the original of which is given in the Appendix to his Geological Researches in China, Mongolia, and Japan ("*Smithsonian Contributions to Knowledge*," vol. XV., Washington, 1867, p. 119). The letter is dated from Cleveland, Ohio, September 25th, 1861. I think the quotations ought to have been expanded, and some words *restored* to what they are in the letter itself. I will, therefore, refer to that document more fully than I did in the last Edition in which I quoted from the report of the Victorian commentator.

Mining Commission and with the authority of the Mining Commission? (A.) I should not consider myself competent to do so; that is all I can say about it.

591. (Q.) With regard to the particular report from which that is extracted, did you ever see the report from which that is an extract? (A.) I never saw it when it was finished.

592. (Q.) I allude to that letter? (A.) I never saw that letter.

638. (Q.) Professor M'Coy in reply to a question states in his examination on the 18th August, with regard to the Cape Paterson Coal-fields :—" That a shaft should be sunk, &c., &c." Are you prepared to state the cost ? * * * * * * besides, there you have the absolute certainty that there are good beds of Coal ? (A.) You see that Professor M'Coy gives evidence about Cape Paterson, but the fact is he has never seen the place. He has never been out of Port Phillip Bay in that direction. The only evidence he gives is from what I described to him about a place. He has never seen the place, so that a person cannot generally give evidence about a place he has never seen. I have walked the coast from the Bass River to Anderson's Inlet, past Cape Paterson, a distance of about 40 or 50 miles.

Frederick M'Coy, Esq., F.G.S., examined, 18th August, 1857 :—

461. (Q.) *By the Chairman.*—You mention the Cape Paterson Coal-fields. Have you any information respecting them ? (A.) Only a report in former years, and specimens from those beds.

462. (Q.) Have you examined them ? (A.) No, I have not. The specimens show them to be identical with the beds of Van Diemen's Land and Sydney.

471. (Q.) The Committee would be glad if you will state from the evidence that presents itself, whether you consider that Cape Paterson Coal-field is most likely to be a large and useful bed for commercial purposes? (A.) Oh! certainly.

Several species of plants are described by Dr. Newberry, and assigned to either a Triassic or a Jurassic age, leaving that age undetermined (from want of sufficient evidence) in a large part of the great Coal-fields of China, basing his "conclusion on the entire absence of Carboniferous plants from the *collection*, and the presence of well-marked Cycads, species of Podozamites and Pterozamites closely allied to if not identical with some heretofore found in Europe and America." He then says—"the Coal basins *you visited* are all Mesozoic, and not Carboniferous." Towards the close of his letter he arranges the plants in four divisions, assigning them all with the exception of one plant to Triassic beds, the exception being one Podozamites, which "*resembles*" a European Jurassic plant, the other apparently being "*identical* with an American Triassic species." There is also a Pecopteris having a remarkable likeness to *P. Whitbyensis*, (which on comparing Pumpelly's figure with those given by Lindley and Hutton and A. Brogniart, I should hesitate to say is actually *identical* with the Scarborough species—though all the figures have some resemblances to each other), and which Mr. P. says is too imperfect to determine accurately. There are other plants, but the balance is with by far the majority, with Triassic beds in Europe, North Carolina, Virginia and Mexico. A few new plants are also mentioned.

When, therefore, such statements are cited to prove the Oolitic or Jurassic character of our New South Wales Coal, we might reasonably expect to find that the prominent plants in our Coal Measures have a place in the Chinese Coal Measures seeing that the latter are brought out in evidence to weigh down all opposition to the preconceived opinion on the subject of age. But *what do we find?* we find the following in the heart of Dr. Newberry's letter.

"We have of course no right to assume from the interesting facts your explorations have brought to light, that no Carboniferous Coal exists in China, for it may very well happen, that as in our own country Coal-seams of economical value, but of different ages, will be found there, at points not greatly removed from each other. But geologists will not fail to be deeply interested in the fact, that so large *portions* of the Coal-basins of China, including beds of both anthracite and bituminous Coal—worked for hundreds of years, probably the oldest mines in the world—are *wholly* excluded from the Carboniferous formation. *So large* a Coal-bearing area, indeed, that when joined to the Triassic, Cretaceous, and Tertiary Coals of North America, they quite overshadow the Carboniferous Coals of Europe and the Mississippi Valley, and suggest the question, whether the name given to the formation which includes the most important European strata has not been somewhat hastily chosen. Another interesting feature in the fossil plants *under consideration* is the re-appearance at the far distant point from which they come of genera so well known in European and American geology, *and the entire absence of the species of Phyllotheca, Glossopteris, &c., which have made the Indian and Australian Coal-floras so puzzling to the palæontologist.* There are fragments of a new generic form—probably a Cycad—in the collection, and some obscure specimens that may represent other plants new to science, but the *Pecopteris, Sphenopteris, Podozamites, Pterozamites, &c.,* have a very familiar look, and their resemblance to well-known forms gives fresh evidence of the monotony of the vegetation of the Globe previous to the introduction of the angiospermous forests of the Cretaceous epoch."

I may be allowed to quote here another extract from Mr. Pumpelly himself on the subject of "Jurassic Coal." He says, on p. 62 of his "*Geological Researches*"—"Were there fossiliferous strata of the Jurassic or Cretaceous ages (*i. e.* in China), their petrifactions would be found in all parts of the empire, used as curiosities and as medicines, as is the case with the fossil Brachiopods and Orthoceratites. *This is important evidence in China* where art is based on the remarkable or rather strange in nature. * * * With regard to the Coal-bearing rocks, I have supposed the Coals to belong to the *same age throughout the empire,* excepting a few which seem from their names to be Tertiary brown Coals."

Now, reconciling the quotations, if we can, from Professor M'Coy's evidence as to the value of Palæozoic Coal, and the inferences of the "*Report of the Victorian Coal Board*" from the letter of Dr. Newberry and the extract from Mr. Pumpelly, what is to be done with another passage in p. 9 of that Report?

In it the Reporter having arranged the order of our New South Wales beds (no doubt, conscientiously enough) after his own idea, says—" If their view be correct, it is not likely that seams of Coal as thick and as persistent as those occurring in the Lower Mesozoic beds of New South Wales will be found in any part of Victoria. It is to be regretted that a geological examination was not made of the Northern Coal-fields, during the many years the Victorian Government maintained a staff of geological surveyors, for the purpose of ascertaining by comparison the position of our beds with all the exactness practicable."

"The value of such evidence as the geologist and the palæontologist can give in such investigations as these is priceless. *They* alone can determine where the practical miner can pursue his explorations with fair chances of success."

Thus speaks out the modern Delphi—but what becomes, after all, of the expectation of the anticipated Mesozoic Coal-beds of Victoria, and what must Mr. Daintree, who was one of the staff spoken of, think of the way in which his success in carrying out the investigation recommended at Stony Creek is rewarded, when that very important work is totally ignored by the Palæontologist of the Survey, by whom all the specimens collected, sent to him by me, were examined, and who now has had his eyes so far opened as to acknowledge that some "Palæozoic" Coal does exist in New South Wales ? *

---

* In reference to the above remark the following passages from "Geological Notes, with Plan and Section, by Richard Daintree, Field Geologist, Victoria," may be properly cited :—

"From Newcastle to Stony Creek is but a short trip, and as these are sections on which Mr. Clarke bases his evidence of the Palæozoic age of part, at least, of the New South Wales Coal-seams, it is one of the necessary pilgrimages of the wandering geologist in search of truth. What I saw there I will state in as few words as possible. I saw three shafts on Mr. Russell's estate—ladder shaft, working shaft, and 200 feet shaft."

He then gives his measurements, which are not material to cite in this place, and goes on—

"When the details of these shafts were first made known by Mr. Clarke, as a proof of the Palæozoic age of the Coal, Spirifers, Fenestella, &c., being found in abundance, and Glossopteris associated with and below the Coal, it was suggested by Professor M'Coy that the data given by Mr. Clarke showed the existence of a fault between 'working' and '200 feet shaft,' and that possibly to this fault the reversion of beds might be due, but the Palæozoic character of the fauna was not called in question.

"This error arose from taking the absolute distance between the shafts (360 feet), instead of the reduced distance to the line of dip 280 feet.

"Referring to the extension of Russell's Coal-seams to the Northern Railway, unfortunately at a point where no marked bed of Russell's series can be absolutely identified" [but at that point may be identified both plants and Marine fossils and traces of Coal in the strata there disturbed], " we have an apparently unbroken series of strata dipping in the same direction, and at about the same angle, as those in Russell's Coal-pits, extending from a point at 19 miles 73 chains from Honeysuckle Flat to 21 miles 37 chains from the same place, the beds furthest to the eastward dipping at a greater angle.

"This affords a thickness (taking the angle of dip at 16 deg.) of 2,365 feet of strata, abounding in fossil fauna from bottom to top—very low down in which Coal-seams with Glossopteris occur.

"Fossils from each of the cuttings on the Railway and from Russell's shafts were procured, that Palæontologists may satisfy themselves of their European parallel.

"If it be admitted that the fauna found in the upper strata of these shafts is Palæozoic, then these Coal-seams at least are Palæozoic, and Glossopteris has a much lower range than has hitherto been assigned to it, except by Mr. Clarke.

"Neither does there seem any reason why Mr. Clarke should not place the Newcastle Coal-seams (his No. 3 Carboniferous group) in the upper portion of this Stony Creek group, no known unconformity existing, since no fauna or flora typical of the Mesozoic period has, I believe, yet been found in the said No. 3.

"This brings me to the consideration of Mr. Clarke's present arrangement of the Carboniferous series of New South Wales :

"*First.*—'Wianamatta' beds, with insignificant Coal-seams, the upper beds of which are the probable equivalents of our Otway, Bellerine, and Wannon beds, in which Glossopteris has not yet been found.

D

As to the fact of changing an opinion on conviction of being wrong, he who so changes is not to be taunted with it unfairly, and I do not advance it except to acknowledge that so far as the Professor has gone he deserves respect and honour for the change. My only complaint is, that he has *not gone far enough;* though after what he and his colleagues announced in the examination above referred to, respecting the sole Mesozoic character of our New South Wales Coals, it is refreshing to find him writing in these terms of the Greta and Anvil Creek Coal-seams,—"The beds from "*k.*" to "*n.*" (referring to his re-arrangement of Mr. Keene's specimens) are clearly the Marine Palæozoic Carboniferous rocks, and *the Coal found with them resembles the Coal of the Southern Coal-fields of Ireland of the same age.*" But he adds— without compunction or authority :—" Neither this collection, nor the sections, nor Mr. Keene's collection in the Melbourne Exhibition, bear out the notion that the Glossopteris and Phyllotheca alternate with the marine Palæozoic shell-beds." Now had

"*Second.*—'Hawkesbury' beds, with insignificant Coal-seams; no Glossopteris. To this series Mr. Clarke refers the Grampian sandstones of Victoria, though Mr. Selwyn places them with No. 4. (By Grampian sandstones I mean the beds constituting the Sierra.)

"*Third.*—'Carboniferous' beds, containing the workable Coal-seams, with Glossopteris, by far the most abundant fossil. In the lower portions of this series four (? five) known Coal-seams are interpolated with strata containing a fauna similar in character to that found in the Carboniferous limestone of Europe.

"*Fourth.*—'Lepidodendron' beds, not associated with Coal-seams, as far as yet known.

"If this arrangement is correct—and my experience as a field geologist is entirely in its favour—it is of great practical value to us in Victoria in the search of workable Coal-seams, &c., &c., ... in the hope of finding the Glossopteris beds. It points unfavourably towards the Tæniopteris and Zamites-bearing beds, which we have hitherto regarded as our Coal-producers, but which as yet have yielded nothing better than the Cape Paterson seams.

"Four thousand feet also of these same beds have been tested by boring in the Bellerine District, and have yielded nothing approaching a workable seam.

* * * * *

"All the facts that we have to guide the field geologist in Victoria, in his search for Clarke's No. 3 Carboniferous beds (containing the workable seams of New South Wales) are these,—that they are very low down in the Carboniferous series; that the lowest beds contain a fauna nearly allied to the Lower Carboniferous of Europe; that Glossopteris is associated with all the Coal-seams, and is the most common and characteristic fossil of the said No. 3. This peculiar fauna or flora has not yet been observed in Victoria."

(From "*Yeoman and Australian Acclimatiser,*" August 29, 1863, No. 100, *published at Melbourne.*)

It will be unnecessary to point out to any unprejudiced reader how Mr. Daintree's "Notes" cited above, known as they must have been to the "Reporters on Coal-fields, Western Port," nearly nine years before, contrast with their lamentation in the year 1872, about the "non-comparison" by *Victorian surveyors* of the position of the Coal-beds in the two Colonies, "with all the exactness practicable."

a visit been paid by him to the localities of Rix's Creek, or to Anvil and to Stony Creeks, or to Mount Wingen, such an assertion would not have required fresh denial from me; and to jump from the Wallsend seam to Rix's Creek and Anvil Creek without any examination of the section of the intermediate localities, or to deny the existence of Glossopteris at those and other places among the Marine beds which are so interpolated, is to do away with the whole merit of such a section as the "Notes" pretend to represent.

Since the date of Mr. Daintree's visit, Mr. C. S. Wilkinson, F.G.S., another first-class member of the same staff of excellent geologists on the late Victorian Survey, has succeeded to the office of Geological Surveyor in New South Wales. It may be sufficient to quote one sentence on his authority: "The collection of fossils from near West Maitland, Greta, and Anvil Creek includes Spirifer, Conularia, Inoceramus, Productus, Fenestella, Bellerophon, Crinoidal stems, &c., obtained from the Upper Marine beds 350 feet above the Anvil Creek Coal-seam, from which seam I collected the specimens now shown, containing the Phyllotheca and Glossopteris *Browniana*" ("*Mineral Exhibits*," from "*Mines and Mineral Statistics of New South Wales*, 1875," p. 133, for Philadelphia Exhibition).

I will quote here an additional testimony to the facts already declared, respecting the interpolation of our Glossopteris Coal in the Marine beds. Mr. Odernheimer in his final report to the Australian Agricultural Company, says:—"The lowest Coal-seam at Wollongong rests on older spirifer sandstone, and is covered by sandstone, with Pachydomus shells and a few Spirifers," (p. 88.)

I have paid more attention, perhaps, to the "Report on the Western Port Coal-fields of 1872," than it deserves; but as it contains specific allusions to myself, and in fact is an attack on the evidence I have conscientiously given on the subject of New South Wales Geology, it is only just to that Colony to show that the conclusions arrived at in that report are "based" as much on personal ignorance respecting our territory, and a pre-determination to disbelieve the statements of men quite as much entitled to be believed as the writers of that report themselves, as on anything else. I am thoroughly persuaded that if such personal investigation on his part had taken place, an old correspondent and assumed friend of my own would not have dealt with my writings as he has done.

The advocates for the Oolitic (or as now called, Mesozoic) age of our Coal plead the cases of Richmond in America, and India, as well as China; Africa is unnoticed. It will be fitting to produce evidence on each head.

China.—Mr. Pumpelly is the only authority quoted by the Victorian Board, who make him to have in 1862–65 found in the Coal-beds *fossils* proving that "those beds are *geologically* of the *same age* as the *Victorian, New South Wales, Tasmanian, and New Zealand beds,"* p. 8, and Professor Newberry is quoted as identifying these fossils as those characteristic of Triassic *or* Jurassic ages." (See *ante*, p. 46.) In the *"Ocean Highways"* for Nov., 1873, Baron von Richthofen says, the Pumpelly observations were only very limited in extent, and his map an hypothetical one made up from native reports, "in which he attempted to exhibit among other data, the distribution of the Coal Measures in China." "The favourable result at which Mr. Pumpelly arrived, in respect to the great extent occupied by Coal-bearing strata in China was modified in some measure by the *somewhat unsatisfactory conclusion* drawn by him, from the determinations by Dr. Newberry of a *few vegetable remains*, that all the Chinese Measures are of the same age as the *Triassic* formation of Europe," (p. 311). The Coal of China, however, found a highly qualified expositor in Baron Von Richthofen himself, who from 1868 to 1872, made journeys nearly all over China, and found Coal-fields of enormous extent in many districts, nearly every one of which he personally visited, as he tells us in various publications.

He mentions one seam of Silurian age; several others in Devonian strata; but he adds *"the great bulk of the most widely distributed and most valuable Coal-beds are proved by numerous and very characteristic Marine fossils to belong to the true Carboniferous.* After the close of that epoch the deposition continued without interruption through the Permian, till probably towards the *close of the Triassic epoch."*

These are his own words, and he justifies his determination of epochs by informing us, that "he first determined with some accuracy the geological age of the Sedimentary formations by a great number of prolific fossiliferous localities." Nowhere in this account of his do we find mention of Oolitic or Jurassic Coal. So that really China should not be quoted to uphold the *"same group as the Cape Paterson series,"* (Report, p. 5). Rather might it uphold the Coal of New South Wales. If Marine fossils are "necessary," none exist in Victoria as we have already seen and as the Report allows.

The following passages from a notice of Richthofen's discoveries concisely meet the facts he had developed, in the Provinces of Liao-tung and Shan-tung:—"Tutti questi strati sono apparentemente quasi paralleli fra di loro, e subiscono soltanto un leggiero cangiamento di inclinazione indicante il graduale passaggio da un livello geologico ad un altro. Sarebbero queste località importantissime a studiarsi, giacchè sembra che vi esista la intiera serie Paleozoica dal Silurico al Carbonifero. Tutta siffatta serie è fortemente disturbata da roccie eruttive, e segnatamente da graniti e da porfidi; la massima intensità di queste eruzioni si verificherebbe nei dintorni di Pechino.

"La formazione Carbonifera di Pechino ha uno sviluppo straordinario." ["*R. Comit. Geolog. d'Italia, Bulletino* 9–10, 1871, p. 234."] "Presso il lago Poyang il deposito scistoso, ora accennato, è ricoperto da regolarissimi strati Carboniferi, fra i quali sono intercalati alcuni straticelli calcarei ricchissimi di brachiopodi in perfetto stato di conservazione. Questa fauna differisce essenzialmente da quella che vedesi associata al carbon fossile nelle provincie nordiche della China: il genere *Productus* vi è prevalente per numero, ma il caratteristico *P. semireticulatus* vi è scarso e rappresentato solo da piccoli individui Rarissimi sono gli esemplari di *Spirifer*, mentre vi abbondano; crinoidi, i coralli, gli spongiarii ed i generi *Orthoceras* e *Porcellia*: sonvi pure rappresentati i generi *Cyrtia, Orthis, Siphonotreta*, &c." [*id.*, p. 236.]

Mr. T. W. Kingsmill confirms these statements in his account of the Geology of the East Coast of China, considering with others that "The Chinese Coal-fields may prove to be the largest in the world, and at a future period will have an important influence on the destinies of the East."*

More recently, in 1873, a letter written to M. Daubrée by M. l'Abbé Armand David states, that in the district of Mien-shien Coal-beds exist in association with Marine Palæozoic fossils and so-called *Secondary plants* which the author describes as *interpolating each other*—"Ce que je ne puis m'expliquer c'est l'existence de ces calcaires durs cristallins *au-dessus* de la houille et *au-dessous*, avec des apparences physiques, identiques, quoiqu'ils soient séparés par 100 ou 200 mètres de marnes." ("*Bull. de la Soc. Géol. de France*," 3 sér., t. ii., 1874, No. 5, p. 406.

He also states that on the mountain of Lèan-chan, near Han-tchong-fou, nearly 3,000 feet high, a grayish-white limestone from 300 to 600 feet thick, having a dip of from 40° to 60°, forms the summit.

Below comes in a series of bluish, red, and yellow marls concordantly stratified with the limestones, followed by red beds like sandstone, the whole system abounding in fossils. Coal occurs above the marl in contact with the upper limestone, which, as well as the shales and clays, contains vegetable and shelly fossils.

Ad. Brongniart describes in the same number of the "*Bulletin*" the plant remains to be *Pecopteris Whitbyensis;* two *Sphenopteris*

* (Notice in "*Geologist*," 6, p. 69, of a paper read before the Geol. Soc. Dublin in 1862. Dub. Q.J.) In a valuable memoir, "*On the Geology of China*," by the same author, we learn that besides Devonian, Marine fossils, and Carboniferous beds containing Lepidodrendon and Sigillaria, and in some places younger conglomerates, and red sandstones not unlike Triassic succeed them, "the Coal, at the latest, being Triassic." In other parts, such as in the Tung-ting system, he tells us that "there is a striking resemblance between it and the Devonian and Subcarboniferous rocks of the South of Ireland—the same succession of grits and shales at the bottom, and a similar development of limestone above, while the type of the few fossils found seems likewise to approach that of the Lower Carboniferous rocks of Europe." (Q.J.G.S., xxv, pp. 119–138, 1866.)

of imperfect character; resemblances to *Zamia distans* and to *Lycopodites Williamsoni;* a probable *Palissya* and *Bayera dichotoma* —the whole very near if not identical with those of the Whitby beds. These come from Tin-Kiako, South Shen-shi. On the other hand M. Paul Fischer describes M. David's Marine specimens from Léan-Chan as consisting :—Firstly, of Crinoidal remains Polyps and Bryozoa and Brachipods, as *Orthis, Ptylodictya, Discopora, Heteropora*. These he considers Wenlock. Secondly, there is also a reddish or whitish sandstone with beautiful fossils such as *Productus*; *Spirifer*; *Euomphalus*; and *Orthoceras*—the rock resembling that of Kildare in Ireland, and Avesnes in France. The coral *Michelinia* and some others were undetermined. "It appears strange," says M. Fischer, "that Upper Silurian and Carboniferous beds should occur together in the same locality."

M. Bayan also describes elsewhere in China, in Yang-Tsee-Kiang, some drifted fossils of true Carboniferous species. Whether the plant-remains do or do not belong to the same beds with the Carboniferous Marine fossils as M. David says, or are altogether younger, at any rate the Carboniferous fossils are Palæozoic, and further researches may demonstrate a more intimate relationship than now appears with the stratification and palæontology of New South Wales. But if there are indications of Mesozoic formations in some parts of China (as shown by Dr. Newberry), yet all observers confirm the fact that the enormously developed Coal Measures are not Mesozoic but Palæozoic Carboniferous. Mr. Pumpelly's view is that all the older Coal over China is Triassic resting on no other Sedimentary rocks, than Devonian. Those fossils of the latter epoch to which he refers I have arranged in the table below, marking those which are known to me to occur also in New South Wales and Tasmania.

There is an interesting passage in "*Siluria*" (4th Edn., 1867, p. 18) which may be properly cited. Sir R. Murchison says therein,—" It is also certain that the mountain-chains of China are composed to a great extent of these older rocks; for M. C. Skatschkof, Director of the Russian Observatory at Pekin, when preparing an account of the rich Coal-fields (partially described by his countryman Kovanko) near that city, recognized, in the Jermyn-street Museum, certain Silurian Graptolites and Orthoceratites, Devonian Spirifers, and Carboniferous Producti, as all being forms which he had seen in the rocks around the Chinese metropolis." He then mentions the fossils given to him by Mr. W. Lockhart (*see* "*Address in R. Geogr. S. J.*," 1858, p. 306) "some from the province of Szechuan and others from Kwangsi," and those brought by Monsr. Itier, and described by De Koninck as Devonian. These are enumerated in the table. But there are others of which at present I cannot refer to a description, nor have I now Richthofen's last work at hand.

PALÆOZOIC FOSSILS, REFERRED TO BY RAPHAEL PUMPELLY, ESQ. 1866.

	Genus.	Species.	Age.	Authority for Species.
...	Aulopora	tubæformis	D	Davidson*
...	Cornulites	epithonia?	D	„ *
...	Spirorbis	omphalodes?	D	„ * } Q.J.G.S.
...	Crania	obsoleta	D	„ * } ix. 1853, p. 353.
...	Cyrtia	Murchisoniana	D	„ *
...	Productus	subaculeatus	D	„ *
...	Rhynconella	Hanburii	D	„ *
...	„	Yuenanensis	D	De Koninck* Bull. Ac. R. Belgique: XIII. pt. 2, p. 415, 1846.
...	Spirifer	Archiaci	D	Murchison* "*Siluria*," 1859, p. 425. 1867, p. 400.
Tas.	„	Cheeliel	D	De Koninck* op. cit.
N.S.W.	„	disjunctus	D	Davidson op. cit.
	„	Vernuilii	D	Murchison* op. cit.
	Terebratula	cuboides	D & C	Guerdet ⎫ Comptes Rendus,
N.S.W.	„	pugnus	D	„ ⎬ Acad. des Sciences.
N.S.W.	„	reticularis	D	„ ⎭ LVIII. No. 19, p. 878.
	Orthoceras		D	Woodward Q.J.G.S. XII. p. 379.

* See also D'Archiac, "*Géol. et Paléont*," 1866, p. 461.

Virginia.—"The Coal Measures of Richmond," says the Western Port Coal Board, "are stated by Sir C. Lyell to belong to the lower part of the Jurassic Group." Well! he did once say so, but he found that he was wrong, and so he placed them finally in the Trias; Professor Heer considering that the plants have "the nearest affinity to the European Keuper." (*"Student's Elements of Geol.,"* 1871, p. 362.)

Why cannot the Board follow a good example and condescend to look down the line a little? They flirt with the word "Mesozoic" out of compassion for their "first love" among the Oolites, and are afraid to acknowledge they have a hankering after a second idea, and so are unjust to it by their indecision.

Africa.—In Africa, the association of the genera Glossopteris, Phyllotheca, and Dictyopteris, "affords some evidence of Mesozoic affinities" says Mr. Tate, who, nevertheless, shows that the shales in which they occur are not Jurassic but Triassic (Q.J.G.S. xxxiii, p. 142.) Palæoniscus and some of the reptiles and an encrinital stem might refer these Karoo beds to a lower position still. Mr. Tate admits the analogy is with the Keuper (p. 169).

The late W. S. Macleay, Esq., F.R.S., always expressed his belief that certain beds near Sydney belonged to some part of the "New Red." And it is curious to observe, how many persons who "know what they are talking about," some from above as the Ooliticals, and some from below as the Permianites and Upper Carboniferites, have found their battle-field on the territory that was once intact as the "New Red," but which has been cut up and re-distributed since the early days of our geological recruiting, after the fashion in political contests. The defenders of the Palæontological territory will not, however, surrender at discretion, but will go in for a final struggle, in the hope and intention of making their case good until they have been *proved* mistaken. It is not so much, however, for the love of the past discussion, as to contribute to the history of it, that in this place, notwithstanding some recent light has been thrown on the Palæo-botany of India by one whose ability and knowledge are deserving of universal respect, that the letter of my friend, Dr. Oldham (published in the last Edition) will find a place in this, for it contains a concise view of what was believed in India by those who used well all their opportunities up to 2nd April, 1874; and if there is error in any of its conclusions we shall have an opportunity further on of comparing the antidote with the bane, and I would hesitate to strike out unceremoniously from these pages the results of years of patient and conscientious labour of one who has "left his mark" upon Indian geology, which cannot be erased without deep ingratitude and deliberate injustice. By comparison of this document with certain revelations to be mentioned in the next section (on the Mesozoic or Secondary formations), the new discoveries will be made plainer and the old rectified where they may have been defective, and I may repeat, in giving a summary of the Indian Coal-fields History as it was about four years ago, I shall, I believe, involve no breach of confidence by doing what will save the necessity of again searching the Memoirs and Records of the Survey:—

"We have seen," he says, "no reason whatever to alter our views with reference to the age of our Indian Coal-rocks. The plant evidence is tolerably conclusive with us. Our *Upper* beds, which contain thin patches and threads of Coal (and which we call RAJMAHAL formation), we have established, by a careful research in Cutch, to be *Upper Oolite*. These are characterized by an abundance of Cycadea and Tæniopteris, but not a single Glossopteris has been found. Then we have the group we call the PANCHET System, with no Cycads. Schizoneura (a plant first described from the Vosges), &c., and with them Labyrinthodont and Dicynodont reptiles. No Glossopteris here either.

"Then below these, with slight unconformity, occur the Coal-rocks, in which, observe, we find *Glossopteris Browniana* abundant; and this holds through the several thousand feet of thickness, occurring in all.

"At the base we have a small thickness (relatively) of the TALCHEER System, in which Cyclopteris shows, but no Glossopteris.

"Unfortunately we have as yet no animal remains from our Coal-rocks. Notwithstanding this, in connection with your evidence from Australia, and bearing in mind the perfectly established identity of the Glossopteris, even in its varieties, and the *equally established fact that Glossopteris has never been found in Europe*, and therefore gives no clue or index to age from European determination, I cannot come to any other conclusion than I have done, that our Coal in India represents *the latest portion of the Carboniferous of Europe, and the gap between this and the Permian;* or, I would say, in a broader sense, *the latest part of the Palæozoic time.*

"I read Daintree's paper with much interest, and think he has done much to clear up some of the difficulties.

"But so long as some fancied analogies with regard to fossils are allowed to sway the mind, there can be no agreement of opinion.

"The Glossopteris of Australia and India are identical. We have every variety, as described from your beds, and no one could hesitate to admit that the beds are similar also. All these Glossopteris beds must be admitted to be of similar relative age in both countries. It proves *nothing* as to the age *relating* to *European systems.* You know better than I do the amount of co-existing evidence as to age which you have established in Australia.

"In India it is this, in a few words:—

(3.) *Above*—A system of rocks, with abundance of Cycads, Tæniopteris, Pecopteris, &c., &c., truly Oolitic with their threads of Coal.

(2.) *Next,* separated by considerable time beds with Schizoneura, Pecopteris (*no* Tæniopteris, *no* Glossopteris), Labyrinthodont, and Dicynodont reptiles, the analogies of which are Permian or certainly Lower Triassic (*no* Coal).

(1.) *Next*—The Coal-rocks, also separated by unconformity, though slight, which have abundance of Glossopteris and also of Schizoneura of different species—as yet no animal remains.

"There are thus three distinct flora with no species common to each. You can draw your own conclusions.—T.O."

In the above remarks of my distinguished friend are some hints that will not fail to be of use in relation to New South Wales, as well as to other parts of Australia, and it is satisfactory to myself to have so much confirmation of my own views. Though it is true that Glossopteris, not being a European plant, does not confer any claim on itself to designate the age of our Coal beds, yet assuredly as it occurs in our Lower Carboniferous beds as well as in the Upper Coal Measures, it does bear on their association with the greatest force, and the two series of beds must be nearly of the same relative age. That age, as pointed out by Dr. Oldham, and as I have all along stated, must be Palæozoic, either on a parallel with some part of the Upper Palæozoics of Europe or occupying a series of beds not represented there.

For the present I content myself with observing that Dr. Ottokar Feistmantel, Palæontologist to the Geological Survey of India, reports the finding of Glossopteris since 1876 in the Rajmahal beds, and that instead of the same species of *Glossopteris* occurring generally in New South Wales and India, in the Damúda beds which are held to be conformable with those of this Colony, he thinks "with great difficulty we may be able to get only one common species" ("*Records, Geol. Sur.* No. 4," 1876, p. 122.) "It seems," he adds, "that the existence of a connection with the Australian is very weak." *

Dr. Feistmantel (4th Nov., 1877) tells me that Glossopteris occurs both in the Panchét and Talchir systems, so that its species must have "a very wide range," going up from the Australian Palæozoic to the Cutch Middle Jurassic. Dr. Oldham had before in 1860 stated as much. (See p. 210 of "*Trans. Roy. Soc. Vict.*," 1860.)

As to the Coal-beds with no Glossopteris, they will go with rocks of a more recent date, and there can be no objection to class them in the age of the Secondary fossils with which they are associated. Professor M'Coy himself admits—"That on mere fragments of leaves or other most imperfect or ambiguous material no generic nor even ordinal characteristics should be founded." ("*Observations on Vegetable Fossils of Auriferous Drifts, by Baron von Mueller*," 1874, p. 14.) But this argument does not apply where fragments even of the same plant occur in

---

* I cannot help alluding in this place to a passage in my letter to Sir H. Barkly, K.C.B. ("*Trans. Roy. Soc. Vict.*," 1860), the publication of which led to a criticism on the part of my opponent, which was not tempered by the "*suaviter in modo*," though in contradiction "*fortiter*" was conspicuous; and which is recalled to my recollection by Dr. Feistmantel's words above,—"*I would not be surprised when the whole deposit of our Carboniferous series shall be made known, if doubts should arise as to the confidence with which some persons speak as to the correlation of the Australian and Indian Coal-beds.*"

two series of beds. Resting on, or passing into each other without a break, they would assuredly show that such beds are intimately related.

If the idea be abandoned (and *there is no real authority for it*) that Glossopteris is an Oolitic plant, and if it be admitted that a fauna has more weight than a flora, and that it is most probable that floral identity never existed during the same epoch at the antipodes of the European Oolitic area, more reasonable will appear the position assigned by me to the New South Wales workable Coal-beds.

Is it more remarkable that *plants* held to be of Mesozoic age in Europe should be found at the Antipodes in a Palæozoic formation, than that usually considered Mesozoic *mollusca* should be found in a similar formation? And the latter is not merely a conjecture but a fact, attested by Palæontologists of eminence. For instance, Münster in 1841 found the three genera *Ammonites, Ceratites,* and *Goniatites* in one and the same bed belonging to the St. Cassian rocks of Austria ; and now we have Dr. Waagan, of the Geological Survey of India, proving to us that the *same three genera* have been found in the same bed together on the Salt Range, in the society of Productæ, Athyris, and other well-known Carboniferous fossils, pointing out that the Ammonites is there a Palæozoic genus, which he places either in the upper part of the Carboniferous, or as Dr. Oldham considers our disputed Coal-beds may be, about the limits of the Permian and Carboniferous formations.

I may also quote here Dr. Feistmantel's words in illustration of the mingling of fossils of distinctive formations : "We have in India the same cases. The genus *Hyperodapedon,* which is yet known in England only from Trias, occurs here in the so-called *Kota Maleri* beds, which are not older than Upper *Lias* ; this *Hyperodapedon* is associated with Ceratodus, also of the kind that mostly occurs in Trias ; the genus Parasuchus, also a Triassic genus, occurs in the same beds, and with all these *Lepidotus* (of Liassic character) is associated ; or what shall we say when we find in exquisitely Carboniferous beds (in the Salt Range) a Ceratites and Ammonites, together with *Productus costatus* and *P. semireticulatus* on one side, and on the other the typically Carboniferous genus *Bellerophon* (in Europe and elsewhere) high up in distinctly Triassic beds, together with numerous Ceratites?" (MS.) *

---

* Dr. Feistmantel, the most patient and critical expounder of Indian Palæobotany that we have yet had, devotes considerable space to the exemplification of similar interchanges in India and South Africa, not only between animal and plant remains, but especially with plant-beds of different stages ("*Records,*" No. IV., 1876, p. 116), and in "Records" (No. 2, p. 29) gives an explanation thus : "In such cases we must only say, the flora of this or that locality (or

Whilst discoveries such as this are being made from time to time, what obstinate pertinacity is it to continue to maintain that even the stereotyped determinations of palæontologists are incapable of amendment. (For Dr. Waagan's description and figures, see "*Memoirs of the Geological Survey of India*," vol. ix, part 2, p. 351. See also Lyell's "*Elements*," 1865, p. 436, and "*Student's Edition*," 1871, p. 358.)

No where in New South Wales has there yet been found in association with the plant-beds any Marine fauna but one which M'Coy and all other Palæontologists admit to be Palæozoic.

The opposition to Glossopteris claiming its descent from Palæozoic times arose from a misinterpretation of facts connected with its appearance in strata from which Marine fossils that prove the age are missing; and thus it got condemned to be Oolitic only, because it is found in company with other plants of whose pedigree no notice is taken. The manner in which such association is sometimes used is anything but logical—"A," it is said, belongs to "B," and "B" belongs to "C," and, therefore, "C" belongs to "A." "D" is not found with "C," therefore, it belongs to neither "A" nor "B."

Moreover, unless it can be proved that every given plant found in different parts of the world had the same instant of existence in all, there must be always uncertainty as to when we may date its epoch. There is also too often a neglect of the conditions of the Strata in which fossils occur, when they are compared with similar fossils in widely separated regions. We know that Coal-plants did not *grow* in the sea, and if they are found bedded among Marine strata it is clear that there we have a guide as to the age, which is only guessed at elsewhere. It would be of use to keep in mind what Oscar Fraas teaches us in his "*Comparison of the Jura Formations with those of Germany and England :*" 1850, as given by Professor Rupert Jones in Q.J.G.S. vii, "*Notices of Memoirs*," pp. 42–83. We must, however, take geology as we find it, till we can arrive at truer conclusions and safer processes than we now possess. The boundaries of the great divisions, Neozoic, Mesozoic, Palæozoic

---

stratum) is of such an age, and was still growing on the coast when already a younger fauna was living in the sea." Does not the case of Glossopteris, &c., in Marine strata prove the same in reverse order—or as contemporaneous with the Marine Palæozoic fossils, and do not both arrangements show how there may be continuous connections from one formation to another, through survivors?

This has now been verified: and, singularly, the number *five* represents the groups (though not precisely the same) in which Glossopteris according to both writers, Oldham and Feistmantel, occurs, the latter naming Cutch (Jubalpúr) Group; Rajmahal Group; Panchét Group; Damúda Series; and Talchir Shales. (*See Hughes:* "*Karanpura Coal-field;*" *Mem. I. Sur.* vii. Pt. 3, pp. 12, 47, 48; 1871.) These I had noted as published.

may yet have to be modified materially, and many changes may yet take place before the geological millennium arrives when fellow-workers will lay aside their prejudices, their animosities, and their inconsistencies.

Calling formations by the names by which they are at present known, we may, nevertheless, admitting possibilities and probabilities of local as well as of general phenomena, go a little further into the *vexata quæstio* of New South Wales Carboniferous peculiarities.

If on other independent grounds the Upper Coal-beds of New South Wales can be treated as Mesozoic, it must still be borne in mind that Glossopteris and other associated plants belong also to a lower group or portion of a continuous series of beds which are strictly Carboniferous; nor must it be overlooked, that in strata supposed to be missing between the two series, which if present would be Permian (or a new formation of which there is no evidence anywhere), the Glossopteris, &c., would in all reasonable probability appear there *in situ*, for it is incredible that in such a continuous succession, those plants would, as it were, *leap over* the whole original mass of deposits between Palæozoic and Mesozoic without leaving any trace of the existence of the genus.

Now, Professor M'Coy in his "Report on some Fossils from Queensland" (14 Sept., 1861), mentioned *Productus calva* and an *Aulosteges* or *Strophalosia* which Mr. A. C. Gregory found on the east of the Mantuan Downs in 1856, and which I submitted to the Professor for examination in 1860, the year I received them (as mentioned in the last Edition, p. 33), and these were held to be Permian. But Mr. Etheridge considers (Q. J. G. S. xxviii., p. 321) that the existence of the Permian requires confirmation; nevertheless, a shell, possibly a *Strophalosia*, is mentioned as having been sent by me to the Daintree Collection, and this also came from the neighbourhood of the Mantuan Downs, in the Nogoa district. It may be suggested, therefore, that there may be an outcrop of Permian in that vicinity; and, if it be so, it ought also to be remembered that Sir T. Mitchell ("*Trop. Aust.*" 1848, p. 240) mentions the occurrence of *Glossopteris Brownii* at the base of Mount Mudge, and other evidences of the Coal-formation, with Coal but a few miles from the Downs. Daintree has mapped the area in question as "Older Coal Measures" (with *Glossopteris*, *Spirifer*, and *Productus*), in an outlying patch of the Comet and Isaacs Coal-fields and the furthest western portion of that portion of that formation. This possible Permian outcrop would be on the Dividing Ranges between eastern, northern, western, and southern waters, and intermediate between the acknowledged "Mesozoic" and the "Metamorphic" regions of the map.

The object of the above references is to suggest, that Glossopteris is a member of a possible Permian outcrop, which has not been yet sufficiently searched for.

However, the force of my argument depends on this—that it is unlikely that plants which occur in Carboniferous strata and in Triassic and Liassic beds (of which more hereafter) should be missing in Permian strata, could the latter be discovered. And this without prejudice to the fact that in other countries Triassic beds are found to surmount the Palæozoic without the intervention of Permian. The latter was held in 1839-'40 by Professor Dana to represent the age of the New South Wales Coal-beds, and in his first publications on the subject up to 1849. In the First Edition of his "*Manual of Geology*" he recalled that opinion (p. 444), stating "that in view of all the facts, it appears probable that the Coal-beds referred to, both in Asia and Australia, represent the Triassic period." But in the Second Edition of 1875, he says (p. 370): The Coal formation of Illawarra and Hunter River, Australia, is probably Permian, as stated by the author in his notes on Australian Geology." ("*Geol. Rep. Wilkes's Expl. Expd.*" 1849.) Thus he returned like a true man to his "first love." But in the First Edition he added: "In the Australian beds there are heterocercal ganoids, and hence the formation cannot be *more recent* than the Triassic," (p. 444). He thus rejected all Oolitic or Jurassic tendencies, and at the same time intimated the existence of a "Carboniferous" flora, saying: "Rev. W. B. Clarke reports true Lepidodendra from the interior of New South Wales—from which it appears that the Carboniferous flora is represented in Australia." This conclusion he also repeated in his Second Edition, in these words: "It exists also and includes workable Coal-beds in China, India, and Australia; but part of the formation in these latter regions may prove to be Permian," (p. 345). *

The occurrence of species in the position assigned to those named above is acknowledged by geologists in other countries. Mr. Lesquereux thus alludes to them in one of his able special Reports, "*On the Fossil Plants of the Cretaceous and Tertiary formations of Kanzas and Nebraska.*" ["*Prelim. Rep. U. S. Geol. Sur. of the Territories, conducted by F. V. Hayden, U. S. Geol.,*" 1870, p. 377.] "Since the first appearance of land vegetation upon the surface of our earth, what we know of it by fossil remains seems to indicate for our country a precedence in time in the development of botanical types. Large trunks of

---

* An excellent illustration of the way in which the succession in one country diverges from that in another, is given by Mons. de Saporta in a review of the "*Carboniferous Flora of the Department of the Loire and the Centre of France*," as described by Mons. Grand 'Eury, in "*Bull. de la Soc. Géol. de France*," t. 57, p. 367, read 19 March, 1877.

coniferous wood are already found in our Devonian Measures, while analogous species are recorded as yet only in the Carboniferous Measures of England. Though the analogy of vegetation between the flora of the Coal Measures of America and Europe is evidently established by a number of identical genera and species, we have, nevertheless, some types, like the *Paleoxiris*, which are considered as characteristic of strata of the European Permian, and which are found in one of our Coal Measures as far down as the first Coal above the millstone grit. Even peculiar ferns of our Upper Coal-strata have a typical analogy with species of the Oolite of England. Our Trias, by the presence of numerous cycadeæ, touches the Jurassic of Europe. But it is especially from our flora of the Lower Cretaceous that we have a vegetable exposition peculiarly at variance with that of Europe at the same epoch, and whose types so much resemble those of the European Tertiary that the evidence of the age of the formation, where the plants have been found, could not be admitted by Palæontologists until after irrefutable proofs of it had been obtained."

If such "seeming discordance" is the case in America, why should not the same view be taken of the relations of the European and Australian Coal Measures? There can be no greater discordance between the relations of the latter than with the examples quoted above, with the additional fact, that in Australia the Upper Coal Measures offer no evidence of any undisputed Mesozoic animal species.

In another place (op. cit., p. 374) the same accomplished author says: "The Lower Permian has still for its land vegetation many species of plants of the Coal Measures, but here the conifers appear represented for the first time by their leaves and branches, and are of a peculiar order. * * * * The Triassic which, with us at least, touches by the character of its flora to the Jurassic, has plants which, like *Cycadeæ*, rather indicate a warm than a vaporous atmosphere. But for this and the following formations, the Jurassic, the data furnished by fossil plants on this continent are too scant to permit reliable conclusions."

What appears to me to be a conclusive opinion has been offered by Dr. Julius Haast, F.R.S., respecting the occurrence of Marine and plant beds of the same age as ours in the Malvern Hill District, Canterbury, New Zealand, who says, in October, 1871 ("*N.Z. Geological Survey Reports on Geological Explorations during 1871–72*"), that on the west side of Mount Potts, Upper Rangitata, there are "different species of Spirifera; besides them there are species of Productus, Murchisonia, Euomphalus, Nucula, Orthis, and Orthoceras. Most of these shells, of which some broad-winged Spirifers are very numerous, are, according to Professor M'Coy, of Melbourne, identical with Australian fossils, and are of

Lower Carboniferous or Upper Devonian age." "Other beds," he adds, " of equal importance occur in the Clent Hills, in which I gathered a rich harvest of fossil ferns, mostly Pecopteris, Tæniopteris, and Camptopteris" (this, however, is not found in New South Wales) "which, according to Professor M'Coy, are of Jurassic age identical with beds belonging to the New South Wales Coal-fields; and although I believe this Clent Hill series to be somewhat younger than the Spirifera beds, I demurred to this definition, owing to the fact that the position of the strata and the character of the rocks of which they are composed have quite a Palæozoic facies."

"Since then it has been shown, and as I think with conclusive evidence, that both fossiliferous strata, the Spirifera and Pecopteris beds, occurring together in the New South Wales Coal-fields, are of the same age, and alternate with each other. The occurrence of Tæniopteris, which hitherto has been considered only of Secondary age,* seems to speak against a Palæozoic origin; however, I may point out that the same objection was made to the Glossopteris in Australia, but which has by overwhelming evidence been shown to be also of Palæozoic age. I do not think that the fragment of a leaf, however distinct, can unsettle all that stratigraphical geology has proved to be correct," (p. 6-7.)

Some recent researches made by me, with a view to the consideration of this question of age, render it far from improbable that a series of beds has been swept off the Coal Measures by denudation, in which Marine beds may have overlain the now existing strata, just as in a lower horizon they do still at Stony Creek, Anvil Creek, Mount Wingen, and in other localities. The facts that the present Coal-seams range in elevation along the coast, from below the sea to between 200 and 300 feet only above it, and that to the westward they reach an elevation of upwards of 3,000 feet, still preserving the same plants as below, and with an equal almost horizontal level (except in cases where local derangement has occurred from special elevating forces), and moreover, that similar seams occur at various other elevations between those mentioned, induce me to consider it possible that there has been a sinking along the coast-line, allowing denudation to operate.

At present this hint may not be worth much, but hereafter more may come out of it. I ought also to add that between the Hawkesbury rocks and the Coal there is often a series of beds belonging to the Coal Measures in which Marine Palæozoic fossils are stated to have been found.

---

* Schimper says (*tome* 1, *p.* 600), of the genus Tæniopteris—"*Ces Fougères paraissent être propres au terrain houiller Supérieur et au Permien,*" i.e., they are Palæozoic. It is only recently that I have obtained not only species of the subgenera, but *real* Tæniopteris from New South Wales, and it is respecting such only that I have written in using the name, in relation to Palæozoic Carboniferous rocks.

In the sections published some years ago by Mr. J. Mackenzie and myself, and in subsequent sections by the former, as given in his Report to Government, it will be seen that the number and thickness of the seams vary considerably in different localities. The former circumstance may be accounted for by the fact that the beds in the Coal Measures since troubled by upheavals, sinkings, and denudation, were deposited over various older formations, some here, some there, which occur at different levels, so that some of the strata are missing in a few of the localities, and all are seldom seen together. Thus the Coal-series at the height of 3,000 feet does not contain so many seams as nearer to the sea level. And, perhaps, in describing them it would be preferable to separate the deposits into various local basins or saucers; though the conditions of a true basin can only be exhibited on the large scale.

It is at least certain, that in the Western Districts, though many of the conditions of the Newcastle and Illawarra beds exist, there are found certain fossils which are not found in the latter, and which would lead to the presumption that, as we ascend in height above the sea we find the introduction of genera gradually approximating to a more recent epoch. For example, the upper beds of the Lithgow Valley Coal Measures contain a fossil which I first collected in 1863, and of which Mr. Wilkinson has lately gathered some striking examples. These coniferous fossils consist of stems and branches ending in Strobilites. Professor Dana, to whom I sent specimens, informed me that he had never seen such in N. S. Wales before. To me they appear not unlike the Strobilites from the *Grès bigarré* of Soulz-les-Bains, in the Vosges, figured by Schimper and Mougeot ("*Monographie des Plantes Fossiles de la Chaine des Vosges*." *Leipzig*, 1844. *tab.* xvi, p. 31.) Dr. Feistmantel considers them as belonging to *Echinostrobus*.

In another direction, viz., on the Clarence River, there is a patch of Coal Measures in which there are forms resembling that of Voltzia, with abundance of fragments of a plant common in the Mont d'Or Coal Measures of New Caledonia, together with plants that have a Tæniopteroid character but are not Tæniopteris, as is the case with many other localities. On the other hand, on Bundanoon Creek, in the County of Camden, there is a Dictyopteris.

As far as some of these plants are concerned, it may be admitted that they are in an unsatisfactory condition at present; but the balance in favour of a " Carboniferous" age for some of the Glossopteris beds is, to my mind, conclusive.

The question, then, about the age of some of the Australian Coal must be considered as settled ; and if, as in Illawarra, the Coal-beds at the base mix with the Marine beds, or immediately overlie them as they do in the Fingal district of Tasmania, it would appear that all these separate occurrences belong to one thick series, in which Marine beds and fresh-water beds interpolate each other. But, assuredly, in that case, the arrangement adopted must express the order as follows:—

1. Upper Coal Measures.
2. Upper Marine Beds.
3. Lower Coal Measures.
4. Lower Marine Beds.

So far as I know, the latter rest frequently on a conglomerate, which in Tasmania I found to contain undoubted Carboniferous fossils.

Hydro-Carbonaceous Shales.

Since the Exhibition of 1862, on which occasion, in a paper on the Coal-fields, I noticed the occurrence of oil-bearing Cannel Coal at the foot of Mount York, and at Colley Creek in the Liverpool Ranges (not on eastern waters), the former has been in great request for the purpose of producing illuminating oils; and the produce has been brought into the market. In the former locality, and in Burragorang, I have made some researches which have satisfied me that these can only belong to the Upper Coal Measures. At Burragorang the blocks of Cannel are found in an intermediate position, between the top of the Coal Measures and the Upper Marine beds, which (if the overlying measures themselves do not) certainly bear the very strongest resemblance to a part of the Hunter River series. (*See Map and Sections.*)

In Illawarra, also, there are Shales which are above that geological position, and which produce oil for illumination, but are not of the peculiar character of the Cannel at Mount York, which in a great degree resembles the Bog Head mineral of Scotland, only it is more valuable. The character of this substance is such as to justify its being considered a species of Bathvillite or Torbanite, in consequence of its colour and woody condition.

It has unquestionably resulted from the local deposition of some resinous wood, and passes generally into ordinary Coal, many portions of the same bed in the Illawarra mines exhibiting the unmistakable features of the latter and the impress of fronds of Glossopteris as plainly as they are shown on ordinary Coal shale. This hydrocarbon varies somewhat in composition; and (as at Colley Creek) is frequently filled with quartzose particles, showing that it was deposited in a shallow pool, to which sand was drifted perhaps by the wind.

At Reedy Creek, now called Petrolia, there is a band of thin and very elastic substance of this kind, separated from the thicker bed below by a parting of white clay.

Varieties of this mineral occur in the Grose River, at Burragorang, on the Colo, on Mount Victoria, and in one spot in Tasmania behind Table Cape, on the southern shore of Bass's Strait, as well as in other localities in other Colonies. Presuming that the origin above suggested is correct, viz., the occasional occurrence in the ancient deposits of trees of a peculiar resinous constitution, there is no anomaly in finding in one spot a mere patch amidst a Coal-seam (as is the case at Anvil Creek, on the Hunter River), or thick-bedded masses of greater area as in the Coal-seams of Mount York, or of American Creek in the Illawarra, depending on the original amount of drift timber.

In the section presented by the escarpment on the left bank of Cox's River, below Pulpit Hill, at Megalong, there are two beds in which this hydrocarbon exists.

Some time since specimens of this, together with others from the Illawarra, were taken to America by Mr. Consul Hall, and were subjected to examination by Professor Silliman. The result was afterwards published in the "*American Journal of Science and Art*," under the name of Wollongongite, an accidental misnomer (as I have elsewhere pointed out), inasmuch as I have Mr. Hall's written assurance that the specimens examined by Professor Silliman did not come from the Illawarra, but from the western sections at Megalong and Reedy Creek.

Professor Silliman shows that this material, as tested by him, has an illuminating power very much greater than any other yet known. It would be invaluable if it existed in sufficient quantity to meet all demands upon it. As it is, there are two separate oil-producing works (one on American Creek, the other in Petrolia), which are now employed in making mineral oils of reasonably good quality, though both inferior to the product described by Professor Silliman.

It has been an object of inquiry whether Petroleum springs exist in New South Wales. Such have been reported from the Corong in South Australia, and from Taranaki in New Zealand, and from Victoria. The former is, we learn, a mistake, being probably at a point where certain animal substances have decomposed. In New South Wales there are also two localities, known to me for many years, in which *a nitrous product* exudes; and there are two or three in Western Australia of the same kind, numerous specimens of which I examined. Nothing of value has as yet been found.

Supposing the truth of the conjecture respecting the formation of Torbanite and its allies from chemical decomposition and changes of resinous kinds of drift timber in the masses now transformed to Coal, the occurrence of such a mineral is not necessarily confined to Coal-beds of one epoch; and thus we find Dr. Hector reporting on the occurrence of a hydrocarbon in New Zealand, from what he deems a Secondary formation, intermediate in volatile matter between those of Torbane Hill and New South Wales, the latter having by far the greatest amount, with much less ash than the former.

Fuller statements respecting the localities may be found in my paper "*On the Occurrence and Geological Position of Oil-bearing Deposits in N.S.W.*," Q.J.G.S. xxii., p. 439. The reader will also find numerous local sections of the Coal-beds in various parts of the same Colony in the "Reports of the Department of Mines," by John Mackenzie, F.G.S., Examiner of Coal-fields, and C. S. Wilkinson, F.G.S., Government Geologist (1875-6), and especially in the work entitled "*Mines and Mineral Statistics*," 1875, prepared for the Philadelphia Exhibition. Mr. Mackenzie has also given sections from what the Victorian authorities call the "*Carbonaceous*" rocks of their Colony.

§ 5. Mesozoic or Secondary Formations.

It must not be supposed, from my strong advocacy of a Palæozoic age for the workable Coal of New South Wales, that I repudiate the existence of the Secondary Formation in Australia; or that, because I oppose an Oolitic or Jurassic age for our Coal-seams, I consider that no Coal, however insignificant it may be, does exist in Australia, or even in New South Wales, which is younger than Palæozoic. There is sufficient evidence in the preceding pages to the contrary to do away with that idea, besides having done my best to bring to light the great Mesozoic formations of Queensland (See various notes by myself in the "*Quarterly Journal*," and the valuable Memoirs of Mr. Daintree, F.G.S., and Mr. Moore, F.G.S., xxvi., 226-261). Although I hold the opinion expressed above, there *are* deposits of Coal of inferior value as relates to extent of area, in Queensland, Victoria, Tasmania, and New South Wales, from which the distinguishing typical plants are excluded; and which, till the discovery of such, must remain, taking into consideration also their stratigraphical position, of more recent age than the rich deposits of the Illawarra, Hunter River, Talbragar, &c. I can only say, that whether I have been mistaken or not in any given case connected with the geological epochs of Australasia, it is not from want of honest devotion to the cause of truth, nor from a desire to hold my own without reason against those who differ from me, that I have in so many publications during more than thirty-eight years of earnest inquiry, defended what I conscientiously believe.

The rule, I think, in such a case as that before us, should be laid down, that plant remains *by themselves* prove very little as to the uncompared age of any formation, but when *associated with Marine fossils, whose age is determinable*, they must go with that formation, of whatever age it may be; for although plants may be swept into the ocean at any period of their existence, they could not be bedded in the same masses of stone formed in the ocean and amidst the Marine fossils, without belonging to the epoch of the latter.

Such is the case in Australia with Glossopteris, and perhaps some others; hence I claim for that at least a Palæozoic age. And so with those described by Mr. Etheridge and Mr. Moore (in the Memoirs above cited) the Mesozoic *Marine* fossils prove the plants to be of that epoch; and when the same plants occur in strata which can be referred to a Secondary formation, and in such also as are Carboniferous, it may be readily granted that they are common to the two. But in the case of Glossopteris no indication is at present producible of its existence in the later formations.

We may therefore refer certain deposits in Queensland, in parts of New South Wales, or the Coal series of Victoria, to Mesozoic (not Oolitic) times, without trenching on the Carboniferous indications. I do not profess to know—and I know no one who is able to tell me—why such arrangements exist (especially as Mr. Carruthers's doctrine is true, for instance, that Tæniopteris and Glossopteris are akin in structure) as place plants very much alike in some respects in different epochs, without confusion, when also the position of the strata is what is called "conformable."

It is no logical argument to say that, because there may be *great* deposits of Coal in China or America or Great Britain that are not what are called *Carboniferous*, therefore there ought to be such, for example, in Victoria, when we all know that they have not been yet found to exist there, or that the same citations would bear out the assertion that the New South Wales workable seams are also Secondary; nor can the adroit alteration of the expression *Oolitic* into *Mesozoic*, prevent our considering that the general term was adopted for the more specific one, because those who used it so were aware that they had made some kind of mistake, and did not like to own it.

Now, there are no *known* Oolitic Marine fossils in all New South Wales; and the Oolitic or Jurassic fossils are of such extent and variety in all countries, wherever the regions in which they occur have been explored, that to put the identity of such formations on a few *plants*, that may after all have no strict claim to decide in the cause, would appear to me a very questionable proceeding.

If, for instance, the fishes found by me in the Gib Tunnel Range, near Nattai, are of a "Triassic or Permian" facies, according to M'Coy, and are Permian according to Egerton and Dana, why should the beds in which they occur be set down as

Oolitic or Jurassic, instead of "Triassic or Permian"? Sir P. Egerton has shown that, with Palæoniscus, occur other genera, closely related to Pygopterus, Acrolepis, and Platysomus, all either Upper Carboniferous or Permian genera in other parts of the world.

Then again, why should the Urosthenes of Dana, from a prominent part of the Newcastle local beds, be left out of the same category?

The view then is, that all these beds, ranging in succession one over the other, and being all as I believe of fresh water origin (for the Hawkesbury rocks contain plants, but no animal remains except fishes), have a common relationship, and yet with no pretext for a Jurassic origin on the score of animal co-existences of that era. When we consider that the fishes alluded to, in connection with Coal and Coal-plants, occur at different altitudes, and are all heterocercal Ganoids, we must conclude that there have been physical disruptions, and that there are gaps in the succession occasioned by following denudation, or that there have been repetitions of strata now only partly traceable.

For instance, the fish beds are at Cockatoo Island 16 feet below the sea; at Sydney less than 100 feet above it; 100 feet at Parramatta; 250 feet above it at Campbelltown; 780 feet above at Redbank near Picton; 1,100 feet on Razorback; 2,360 feet at the Gib Tunnel; and 3,450 feet on the Blue Mountains; the lowest two stations and the highest being in the Hawkesbury series, and the others in the Winnamatta beds above the Hawkesbury; whilst at Newcastle, the Urosthenes was the deepest below the sea, and the oldest in position.

As necessary to explain still further the succession of strata, I introduce here some additional remarks on the Supra-Carboniferous rocks in the province of New South Wales.

Hawkesbury Rocks.—Over the uppermost workable Coal Measures, which are of considerable thickness, is deposited a series of beds of sandstone, shale, and conglomerate, oftentimes concretionary in structure and very thick-bedded, varying in composition, with occasional false-bedding, deeply excavated, and so forming deep ravines with lofty escarpments,—to the upper part of which series I have given the name of Hawkesbury rocks, owing to their great development along the course of the river-basin of that name. These beds are not less in some places than from 800 to 1,000 feet in thickness, containing patches of shale, occasionally with fishes, with fragments of fronds and stems of ferns, a few pebbles of porphyry, granite, mica, and other quartziferous slates, and assume in surface outline the appearance of granite, from the materials of which and associated old deposits they must in part have been derived. On the summit of the

Blue Mountains, and along the Grose River, the thickness of the series is very much greater than near the sea. Patches of very small area contain bits of Coal, carbonate of iron, and sometimes represent miniature Coal Measures.

Towards the base, bands of purple shales are frequent, and ferriferous veins, with specular iron, hæmatite, ilmenite, graphite, and other minerals, sometimes occur.

In places, as about the "Yellow rock" near the Upper Wollombi River, in Ben Bullen, and above the deep excavation of the Capertee amphitheatre, salt and alum are found in cavities formed by decomposition; and in other places, as at Bundanoon Creek in the Shoalhaven District, at Appin, and on the Bullai escarpment of the Illawarra, and at Pittwater, north of Sydney, stalactites have been formed under similar circumstances.

There is an enormous mass of brown iron ore highly carbonised, partly worked at Fitzroy, near Nattai, another on Brisbane Water, and a smaller, on the coast, a few miles north of Sydney, and other similar patches in intermediate localities. These are in part associated with specular iron, which occasionally lines the joints of the sandstones close at hand with well-formed crystals.

The uppermost beds of this formation, especially where they become conglomerates, exhibit isolated summits imitating ruined castles, and have thus been traced by me at intervals all along the escarpments to the westward of Sydney, from the latitude of the Clyde River to that of the Talbragar, and in certain localities within the longitudes of that line and the coast. In the deep ravines of the Grose and Dargan's Creek, the one eastward and the other westward of the Darling Causeway traversed by the Western Railway Line, the slopes are studded by fantastic pillars sculptured by denudation and decay into imitative architectural forms. Similar forms cap the extension of the coast range to the head of the Goulburn River. The tints are *poikilitic*, darkening from exposure, and exhibiting imitations of landscapes sometimes of striking character. The semi-crystalline fragments of quartz, and the disposal of colours (suggesting the idea of the action of gases removing the ferruginous tint in places) have caused me to believe that some transmuting agency has affected large areas of the Hawkesbury rocks. The glistening of the crystalline quartz particles reminds one of the same character observable in the millstone grit of England. It is impossible to understand how considerable masses of the sandstones could have received such a present structure without the metamorphism suggested; for the crystalline facets are quite unabraded and belong to particles that have been collected originally by water holding silica in solution. By washing in acids the colouring matter of the particles may be entirely removed, and then it is seen that they are imperfect crystals.

But the cementing matter is not always ferruginous; a felspathic cement holds them together with *used* mica evidently derivative, and sometimes with graphite.

Another variation in character of the Hawkesbury rocks is in their cohesion. In 1850 I was Chairman of the Artesian Well Board, and remember the difficulty we had in procuring tools hard enough to pierce the quartzose sandstone at the gaol in Sydney; the boring after a small depth was abandoned—one of the workmen precipitating the conclusion by blocking the borehole. But in parts of the Railway Lines, there have been instances, as stated to me by the Engineer-in-Chief, when the largest blocks have been shivered to atoms by a not very heavy fall over an embankment.

This group of Hawkesbury rocks has been by some persons denominated "Sydney Sandstone." The designation was derived from the early settlers, who had not gone far into the country; but it is a misnomer, for it neither represents the whole of the series nor the whole of the material of the rocks, besides making confusion with the "Sydney Sandstone" of the Cape Breton Coal-field of British America. The latter has a clearer right, perhaps, to the title of "Carboniferous," as it is of the age of the very lowest of our Australian Coal Measures. Yet with its Lepidodendra, &c., it has fishes of the one genus which occurs in our Hawkesbury and Wianamatta beds, over our Upper Coal-seams.

Wianamatta Beds.—The Hawkesbury rocks are succeeded by another group or series of strata named by me from the Wianamatta, or South Creek, which runs longitudinally through the basin which fills in the area between a surrounding enclosure of the former series which must have been broken up in part and denuded, either completely before or during the deposit of the sandstones over-lying the Coal Measures. The deep ravines which mark the Hawkesbury rocks give place to rounded smooth undulating softer argillaceous strata, in the bottom of the creeks of which and in the beds of the river Nepean or Hawkesbury and of George's River are marks of old erosion in the harder rocks below the argillaceous shales. Pot-holes are very common in the Hawkesbury beds under the Wianamatta strata where exposed at the points of junction at some distance from the present creeks and drainage channels. Such may be traced at Myrtle Creek, near Picton, and on the Windsor Road near Parramatta. These certainly prove a partial or general erosion before the whole series of the Wianamatta strata were laid down. The nearest beds of the latter to the underlying Hawkesbury rocks, are shales which have occasionally filled in hollows previously existing, or contributed patches forming considerable masses as well as thin layers to the uppermost Hawkesbury rocks. In this way fishes have been found at various levels in shale patches, as on the Blue Mountains, at Parramatta, at Biloela (or Cockatoo) Island, and other places near Sydney. The Wianamatta beds are, however, not all shale, for there are fine sandstones more compact and heavier than the Hawkesbury, calcareous sandstones and ferruginous nodules, bearing fishes and small fresh-water molluscs which remind one of the somewhat similar nodules of Permian beds of Germany.

Could I have procured the remains of fishes that have been reported to me from beds below the Upper Coal, and of the finding of which there is pretty good evidence, we might have been able to show that the same genera that we find ranging from the Wianamatta down to the Coal Measures of Newcastle, all through the Hawkesbury series, occur still lower.

A Palæoniscus, found since my discovery in 1860, was exhibited by the Surveyor General (who gleaned after my harvest), in the Exhibition of 1875 at Sydney; and a specimen of Cleithrolepis found in a Railway cutting on the Blue Mountains was shown by Mr. T. Brown, to whom it had been given by the finder after I had had it photographed. These formed part of the collection exhibited by the Mining Department at Philadelphia.

There are in the Wianamatta Beds in places columnar and pisolitic iron ore, abundance of fossilized wood, plant impressions, and calcareous sandstones, which latter form the highest levels and summits of insulated hills that attain but moderate elevation (1100–1300 feet) in the centre or on the outskirts of the basin, which latter is chiefly confined to the heart of the County of Cumberland and part of Camden, of which Bulbunmatta or Razor Back Range and Menangle Sugar Loaf are outlying relics of a once wider extended plateau. Fossil plants abound in some of the shales and fine sandstones, and the whole area is marked either by old trappean or more recent basaltic rocks, which have produced some effects on the beds traversed by them. Very small patches of Coal occur, but no seams nor any of value have been met with. The old Diorite hill of Waimalee, or Prospect, near Parramatta, must have existed long before the infilling of this basin, as the Wianamatta plant-beds on the flanks of the hill have evidently derived their matrix from the Diorite, and have since been intruded into by what is probably Tertiary basalt. Felspathic trap is common in the basin, and may have been connected with this outburst of igneous eruptions which probably formed many of the solitary hills of a portion of the County of Camden.

Victorian Palæontologists claim for that Colony the existence of a Coal formation of the same age as the Wianamatta, and I have myself long ago pointed out that certain beds at the Barra-

bool Hills resemble very closely certain strata about Camden
in New South Wales. But if the latter are proved to be of
younger age than that which has been assumed for them, it is
not necessary to place the two series (so widely separate in
space) on the same actual horizon.

We have not recognized in New South Wales the Cycadeous
plants of Victoria, nor is there a perfect agreement in the
phytology of the Wianamatta and Victorian strata. In 1861 I
mentioned ("*Recent Geological Discoveries*, &c.," p. 45) three of
M'Coy's New South Wales Plants, *Gleichenites odontopteroides*
(called Pecopteris by Morris and Carruthers) ; *Odontopteris
microphylla;* and *Pecopteris tenuifolia*, as occurring in the Wianamatta beds. These are not reported from Victoria, whilst *Sphenopteris alata (Brong.), Grandini of Goepp. and Schimper*, from Newcastle, belongs to the Old Carboniferous in Germany, and not to
any Mesozoic formation.

In the list given in "*Progress Report of Victoria*, 1874," Professor M'Coy mentions 3 species of *Gangamopteris*, from his
Upper Carbonaceous beds ; 2 Neuropteris, 1 Pecopteris, 3
Sphenopteris, 1 Tæniopteris, with 3 Zamites and 1 Phyllotheca
from the Lower Carbonaceous ; and only one animal form, *Unio
Dacombii.* The alleged abundance and value of Coal in these beds
have been proved to be a myth. There is, however, more Coal
therein than in the smaller area of the Wianamatta and Hawkesbury rocks ; and probably that is the reason why the Professor
would place them *below* the former group of New South Wales.
But when we consider the great improbability that a series of
strata having a thickness of at least 5,000 feet could ever have
existed between the Hawkesbury and Wianamatta series, and
that not a trace remains anywhere in New South Wales of such
interpolation,—that the fossil evidence is in opposition to it,—
and that the areas are totally disproportionate,—it would appear
a mere caprice of fancy to hold such a notion as that hinted at.

It may be well to make a final remark respecting Mr. Brough
Smyth's idea, that the Coal-beds of New South Wales lie on
"*limestone*," ("*Progress Report*," p. 26.) Had he visited them
himself he would have seen that limestone, as such, is rather a
rare rock in connection with the New South Wales deposits of
Coal, which clearly interpolates the Marine beds ; but the latter
are more frequently conglomerates, or sandstones and grits.
The Upper Coal Measures rest frequently on granite and slates
as well as on other rocks. The limestones in the Carboniferous
rocks are rare, being few and of limited extent and far between.
The author just mentioned considers the relation of the "*Coal-
bearing*" to "*Palæozoic rocks*" as "obscure," but it is not obscure
to those who have examined for themselves, nor more so than the
feeling which induces *philosophers* to keep out of sight and

ignore the evidence which contradicts their own preconceived opinions. It will have been seen in the preceding remarks, that I have myself suggested the possibility of some part of our Sedimentary deposits having a relation to the Trias, and it is only fair to state that Professor M'Coy, in his earlier writings, limited the New South Wales deposits to the Oolite of Scarborough. Afterwards the term "Mesozoic" was introduced to define the period—which, of course, left all undefined from the base of the Tertiaries to that of the Trias; the limit of range allotted by him for the Queensland Secondary Marine fossils and plants being from the "lower part of the great Oolite" to "the base of the Trias" ("*Annals*," ix, No. 50, Feby., 1862), placing all on one geological platform. He might, therefore, admit the Trias to represent the "Mesozoic" instead of the Oolite, but he has stuck firmly to the latter.

On the other hand, I presumed to think that the plant-beds in the separate Colonies did not represent the Oolite—and that for reasons assigned I have ventured to believe that the Coal-beds belong to some part of the Upper Palæozoic, either represented by visible examples elsewhere, or belonging to strata not yet found elsewhere represented.

In working out my own conclusions, I had recourse to what was reported by the Geological Survey of India; and relying on the data proclaimed, I held that the nearest ally to our Coal-beds was the Damúda division of what is now known as the "Gondwána series or system" of Feistmantel, ("*Records*," No. 2, 1876, page 28.) Dr. Oldham agreed with me, as has been seen; but Dr. Feistmantel having obtained or discovered the existence of plants in beds not previously recognized by the Survey, *e.g.* *Glossopteris*,—and finding it not only in the Damúda, but in the Talchir strata below, and also in all the intermediate strata, through the Panchét up to the Rajmahal and Cutch (Dr. Oldham agreeing with him as to Panchét being Triassic),—has come to the following arrangement:—

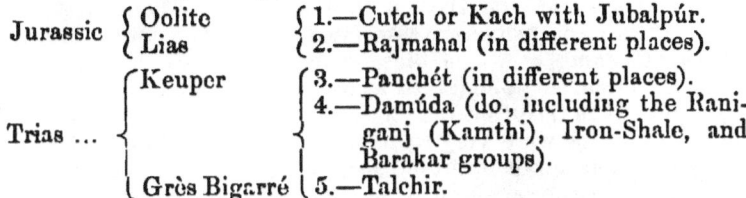

These are the "five horizons" of the Gondwána system.

If this arrangement be correct, then it is clear that there is a a Triassic series amidst the beds which M'Coy held to be altogether Oolitic, but which I, in common with Dr. Oldham, considered in part Palæozoic; and here I may quote Mr. W. T.

Blanford's able Report on the Raniganj Coal-field. ("*Memoirs Geol. Sur. India,*" Vol. iii., part 7, chap. vi., page 135, 1861), in which the author writes in relation to the Panchét group: "So far as this evidence goes, it tends to confirm Dr. Oldham's suggestions as to the Damúdas being Upper Palæozoic. For Labyrinthodont reptiles (and consequently the Panchéts if equivalent to the Mangális) being Permian or Triassic, and the Damúdas being but little older, would be Upper Carboniferous or Permian, or perhaps intermediate between Permian and Triassic; but the evidence is very slight."

It is not strange that sixteen years continued exploration, and a critical examination of the fossils, should lead to a modification of views, and it would be held presumptuous in anyone not on the spot to dogmatise to the contrary.

The present able Superintendent of the Geological Survey in India, Mr. H. B. Medlicott, F.R.S., thus speaks of Dr. Feistmantel's botanical researches:—"Palæontologists come from their cabinets in Europe with the fixed idea that the 'laws' they have seen to work so neatly as between Bohemia and Bavaria, or from Durham to Dorsetshire, will apply equally well between India and Australia, or Europe; and the eager aim of their labours seems to be to tally off our Indian rock-groups as the representatives or equivalents of certain fossiliferous series of Europe or elsewhere. From the beginning this Palæontological fallacy has been a chief obstruction to our knowledge. When first the Gondwána fossils were taken up, pure Geology being in the ascendant, the fact that certain plant forms of the lower Gondwána rocks were somehow associated with beds having a Carboniferous Marine fauna in Australia, was made the basis of a special pleading to show that the Damúdas, their flora, and their Coal, were Palæozoic. The materials have now come into the hands of a pure palæontologist. He has shown, I believe conclusively, that the Gondwána flora is wholly Mesozoic, nailing its several phases to certain representative zones in Europe. But it so happens that on the confines of India, east and west, the upper Gondwána groups are associated with beds having a Marine fauna according to which these said groups have already been attached by palæontological experts to other standard groups in Europe. It is true that the study of this fauna was only partial; but the experts were very accomplished in their line, and their judgment was quite unprejudiced, so that it must carry great weight. Here then again is an opening for the procrustean method of research: and there are symptoms that it is to be duly applied; *this time to make the fauna conform to the flora*......No theologian could be more impious in reducing the mysteries of existence to the compass of his narrow thoughts, than are often scientific specialists in imposing crude conceptions upon the proceedings of nature. Yet these ought to know better—that truth is discovered, not invented." "It is fiction to assume that Palæozoic and Mesozoic faunas have not co-existed upon the earth." "The facts of our Gondwána rocks are certainly puzzling to systematists: on the West in Kach (Cutch) we have the flora of the top Gondwána group, which has a Bathonian facies, associated with Marine fossils of Tithonian affinities; while on the the South-east, in Trichinopoli, beds with a flora (so far as known) like that of the Rajmahal group, which is taken to be Liassic, have been described by Mr. H. F. Blanford ("*Mem. Geol. Sur.*," vol. iv., p. 47) as overlaid, in very close relation, by the Ootatoor group, the fauna of which has been declared, upon very full evidence, to have a Cenomanien facies."

"These questions of homotaxis concern the whole body of naturalists as much as they do us; and I hope some guiding spirits amongst them will keep a watch on our proceedings. Happily these foreign relations do not interfere with the local regulation of our rock systems. The terrestrial fauna and flora of the Gondwánas is developing into a compact unity of its own, and its relations to contiguous Marine fossil faunas is normal, so far as this word can be legitimately used." ("*Records Geol. Sur. India*," vol. x., pt. 1, 1877, pp. 2, 3.)

On this plan of dealing with the Australian and Indian Coal formations in relation to their fossil vegetables, there must be a necessity for very close examination of the individual plants to satisfy the inquiry; and relationship to the Marine fossils of the former which do not exist in the latter cannot, on the plea expressed by Medlicott, be excluded. We may, therefore, look to Dr. Feistmantel's comparison of the respective floras of the two countries.

In order to assist in this I had placed in the hands of my friend Dr. Oldham, at Calcutta, a considerable series of our New South Wales plants, from Newcastle, Hunter River, Illawarra, Lithgow, Merigang, Clarence River, &c., in New South Wales, and from Queensland, Tasmania, and Victoria, including the Coal-seams of this Colony and the overlying strata of the Hawkesbury and Wianamatta groups. A few of these were contributed by Mr. C. S. Wilkinson, and the collection was left in the hands of Dr. Feistmantel when Dr. Oldham retired from the direction of the Indian Survey.

On these, the latter gentleman has reported to me his opinion. Since then, I forwarded at his desire a further collection, comprising additional specimens of plants from the lower Coal-seams of the Hunter, collected in part by Mr. Mackenzie, from Greta, Stroud, Queensland, &c., including the fishes of the Gib Tunnel, with sections and papers, &c. These arrived safely, and an account of them will be found in *Appendix XX*.

This is named to show that no pains have been spared by me to put as full evidence before Dr. Feistmantel of the plants, as I had already done respecting the Marine faunas of the Silurian, Devonian, and Carboniferous beds.

Now, Dr. Feistmantel states that so far from the plants of the Coal-seams being exactly the same as those of the Indian Coal plant-beds, he thinks that—("*Rec. Geol. Sur. India*," ix., pt. 4, 1876, p. 121)*—" those palæontologists who declared the whole Australian flora as absolutely Jurassic ["or altogether Mesozoic," as interpolated in MS. by the author in author's presentation copy], did not distinguish the *Lower* and *Upper* portion of the Coal Measures. The first contains forms which could never support this assertion, while the Upper Measures contain, besides those plants without analogy, some other forms which certainly can justify the supposition of a Jurassic age [altered in MS. as above, to—" Triassic age, some perhaps also Jurassic (Queensland) "]. .

"On page 83, Mr. Blanford gave a scheme of the formations in N.S.W. Coal-fields (1, 2, 3, 4, 5, 6). Nos. 1 and 2 (Wianamatta and Hawkesbury beds), it is true, have yielded no distinct *Glossopteris;* but in Tasmania, from which identical fossils with those of these two beds are known, *Glossopteris* occurs, with *Pecopteris Australis*, *Phyllotheca*, and the most important with *Tæniopteris Daintreei* (M'Coy), (M'Coy : " *Prodrom, Decade II.*," p. 15, "*Rep. Prog. Geol. Sur. Vict.*, 1874, p. 25"). As to 3 and 4, of which the first are the Upper Coal Measures of Newcastle, Mr. Blanford himself (p. 83) says : " Nos. 3 and 4 appear to be connected by the presence of *Glossopteris Browniana* in both, although there *appears to be a considerable distinction in the flora";* and I would add, No. 3 does not contain any animals, while in No. 4 Marine animals are found abundantly.

"On p. 84, Mr. B. enumerates the species, which, as he considers, are common to our Damúdas and the Australian beds, and others which are common to the Damúdas, and the Triassic rocks in Europe (as I pointed out). On these I would remark,—

"*Glossopteris* [" 2 or 3 species identical.—W.T.B."] I think with great difficulty we may be able to get only one common species.

---

* On the question—What is the analogy of our Damúda Series with the *Lower* Coal Measures of Australia? After saying that the analogy is by no means what Mr. Blanford seems to think, he proceeds,—"Any instructive or conclusive comparison ["of our Damúdas" in MS.] can only be made between ["with" in MS.] series that possess fairly represented and characteristic flora. For our Damúdas this condition can only be said to exist in the *Upper* Coal Measures in Australia, and in some exclusively *plant-bearing* rocks of Europe."

"*Gangamopteris* ["The genus only.—W.T.B."] This form is not known at all from those beds intercalated with Marine fossils, but from really Mesozoic beds in Victoria, associated with *Tæniopteris Daintreei*, M'Coy.

"*Vertebraria* ["One species identical W.T.B."] There is as yet no full description of the Australian *Vertebraria*, and that which is known seems to be quite different from ours. The greatest portion of our Damúda *Vertebraria* are probably not identical with those from Australia.

"*Pecopteris* (Alethopteris) ["One species, probably identical, W.T.B."] I doubt whether our *A. Lindleyana* can be united with *A. Australis*, M'Coy, or if this is altogether the case with any other species.

"Thus it seems that the evidence of a connection with the Australian Coal Measures is very weak, while the fossils enumerated as common with European Trias are unmistakably identical.

"*As to the stratigraphy of the Australian Coal strata*, the literature is not poor, but yet it is not in all points quite clear and always trustworthy.

"It is well known that there can be a complete concordance in the stratification of rocks, and yet two or more different formations may be represented which can only be distinguished by the prevailing fossil forms.* As an instance I can quote the Salt Range in India, where, as Mr. Wynne tells us, the Lower Marine Carboniferous and the Triassic rocks are conformably deposited, and yet they are different in age, although a well-marked *Ceratites* and *Phyllotheca* go down into the Carboniferous rocks, and marked forms of *Bellerophon* survived into the Trias. The same relations will have to be applied to the two portions of the Australian Coal Measures, only that here the case is illustrated in the flora.

"For the stratigraphical grouping of the Coal-strata of New South Wales, we must especially take the Rev. W. B. Clarke's observations, which to a great extent are published ("*Remarks, &c.*," 1875); partly Mr. Clarke communicated them to me in two letters; and he sent also a suite of fossils for comparison. From all his *clear* communications it is plain that there are two very distinct portions in the Australian Coal Measures :—

"*a.—Upper Coal Measures.*
"*b.—Lower Coal Measures.*

"*a.*—The Upper portion is marked by a flora which is abundant. Nos. 1, 2, 3 of Mr. B's list must be referred to this; they contain no Marine fossils to indicate a connection with the lower portion.

---

* See a very remarkable instance of this referred to by me, at p. 31, of the position of the Palæozoic formations at a locality in Spain, described by Casiano de Prado.—W.B.C.

"*b.*—The Lower Coal Measures are marked by two Marine faunas of, as generally taken, a Carboniferous age, which separate distinctly these from the Upper beds. The flora is, as both Mr. Clarke and Mr. Daintree state, only rare.

"*c.*—Below this there are beds with real Lower Carboniferous plants."

Dr. Feistmantel then gives the succession of the several strata as I had communicated it to him in a table, and after it a list of plants which he "has seen, or which are mentioned as really occurring," viz. :—

"*a.—Upper Coal Measures.*

" (1.) From Queensland.
" (2.) Tasmania.
" (3.) Victoria.
" (4.) From the Wianamatta and Hawkesbury, we have mostly *Dichopteris, Thinnfeldia, Pecopteris odontopteroides,* Morr., *Tæniopteris,* &c.; and in both the same genus of a fish.
" (5.) From the Clarence River District.—*Tæniopteris* with narrow leaves, and a coniferous branch, which Mr. Clarke himself marked (?) *Voltzia.*
" (6.) Bowenfells and Newcastle.—Here the flora is mostly developed; *Vertebraria,* real *Phyllotheca,* many *Glossopteris* (but few identical with those of India), mostly *Gloss. Browniana,* Bgt., coniferous plants near the Mesozoic *Echinostrobus,* coniferous seed-vessels and others, but *no animal fossils,* nor Lower Carboniferous plants.

"*b.—Lower Coal Measures.*

"I have seen *Tæniopteris,* near *Tæn. Eckardi,* Germ., *Glossopteris,* small specimens; besides these there are quoted *Phyllotheca* and *Næggerathia.* With these are associated Carboniferous (in M.S. 'animal') fossils.

"*c.—Strata below—*With *Cyclostigma Kiltorkanum,* Haught., *Rhacopteris, Sphenophyllum* (real Palæozoic form). These I have seen myself. And again a Palæozoic (Carboniferous) fauna.

"From this we see the following: Only the strata *sub b* can claim a Palæozoic age, containing a prevailingly Carboniferous fauna, which already in *c* occurs together with a palæozoic flora. The flora in *b* is very poor, containing only few forms, which (see remarks p. 165) are so frequent in the upper strata; and to use Mr. Clarke's own words about the *Glossopteris,* we may say,—
'There (in the Australian Lower Coal-beds) it clearly does not

govern, but must be subordinate to the fauna'; and further he says, 'Why might it (*Glossopteris*) not pass into Secondary rocks without denying its existence in the Australian Lower Coal-measures'?" [In MS. he adds, "What I completely adopt."]

At p. 125, he says—"That the Upper beds in Australia—Wianamatta, Hawkesbury—and the Upper Newcastle Coal-beds form a connected series, is also shown by the occurrence of the same fish, which is not found in the Lower strata.

"The following table may illustrate the relations :—

Europe.	Lower Gondwánas, India.	Coal Measures in Australia.
Rhætic ...... } Upper Keuper ... } Trias Grès bigarré } Lower Bunt. sanst. } Trias	Panchét group— Flora and Reptilia. Damúda group— Flora only.	*a. Upper Coal Measures.* All the strata as I enumerated them above under 1, 2, 3, 4, 5, 6. *Flora only.*
Carboniferous	....	*b. Lower Coal Measures.* [In MS. he adds Plants and Carboniferous fauna.]
Carboniferous Devonian ?	.... " Records," ix, p. 125.	*Strata below.* *Goonoo Goonoo.*

In relation to these printed notices, Dr. Feistmantel writes privately,—" Glossopteris began to live rarely in *Australia*, during the time when Carboniferous animals lived in the sea—in the time of the lower Australian beds. They are, therefore, of Carboniferous age. But *Glossopteris* continued to live when already the *Lower* beds were deposited (including the Marine animals), or when the Marine animals ceased to live—when therefore, in fact, another epoch of life began which was characterised by the total absence of Marine Carboniferous animals and by the preponderance of plants; and I think in this lies the difference between your *Upper* and *Lower* Coal-beds, of which only the latter can be considered of *Marine* origin, as *Marine* beds, while the Upper ones are certainly not *Marine* beds. And from this reason, I thought, only those *Upper Coal strata* in your country can be compared with our Talchir-Damúda beds, as these do not contain any Marine fossil at all, and the flora they contain bears a complete *Triassic* facies, so that I do not see any reason why these beds should not represent the Triassic epoch: as for another epoch there is not the least indication. And now, judging from this, I was also convinced that your Upper beds cannot be Oolitic, or even Liassic (except, perhaps, some in Queensland), as they are equivalent to our Damúda series."
(MS., letter 20/2/77).

Dr. Feistmantel has since favoured me with several letters, which relate to a fuller expression of our Australian formations, to which I cannot do full justice in my limited space. I must, however, quote a passage from some remarks of my own, during my discussion with Professor M'Coy, which were read before the Royal Society of Victoria, December 10, 1860, in order to show that any proposal to give a more recent age than that I defended to our N. S. Wales Coal-seams does not take me by surprise: "To sum up all, I may here state that though it is very easy to make the 'worse appear the better reason,' I have no object in any controversy on this question but truth. Having since my acquaintance with the whole of the facts always found a difficulty in reconciling the idea of two epochs in the formation of the deposits including our Coal-beds, in consequence of the apparent continuous succession of those deposits and the occurrence of Coal throughout, together with the absence of Oolitic zoological and the presence of Palæozoic zoological forms, I have not seen fit to renounce the opinion which is shared by others as well by myself, because at present we have no grounds to do so. But it is easy to gather from this paper, as well as from other evidence of my own, that I am quite ready to admit, when proved, that some of the beds are younger than my fourth division, or Mr. M'Coy's base of the Carboniferous system, and may with the example of India before us be even younger than Oolite ; but with the idea of one succession, I must renounce the idea of all above the base being Oolitic."

Future researches may be needed to ascertain—what is possible —that true Palæozoic Marine fossils may be yet detected, in some at present obscure localities, above the horizon of the Upper Coal-seams of New South Wales—or below the base of the present Talchir of India ; and in either case there would be a probability of reversing a decision respecting the claim of the lowest Mesozoic, a contingency which would not take even Dr. Feistmantel by surprise, as he suggested to me (August, 1877) : "The Indian Damúda series you may be pretty certain must turn out *Trias*, or at the utmost *Uppermost Permian* as passage bed between this formation and the *Trias*, but there is nothing as yet which would prove for this, while all is in favour of *Trias*." We must bear in mind, however, a suggession of Mr. Carruthers, that the Permian vegetation shows Mesozoic affinities, and that in fact the commencement of the Mesozoic flora is to be sought in the Permian. (Q. J. G. S., xxv, p. 158.)

The above references under the head of Mesozoic, though alluding to the Carboniferous, are rightly introduced—but not with the intention of accepting the former age as comprehending the latter till further proof has been afforded.

The differences between the conditions of the Damúda Coal-beds of India and the Coal-beds of New South Wales are by the allowance of Dr. Feistmantel sufficiently striking to justify delay until further evidences have been produced for their union or identity. At one time I held the opinion that the Damúda was our nearest ally in the defence of a Carboniferous age, but as that has now to be regarded in a different light, the differences alluded to must be taken to imply somewhat different positions for the two formations. For the absence of Marine fossils in India, and the dissimilarity of botanical species among the plants, with some other particulars, leave a margin for the adoption of a provisional later date for the one than for the other.

But I say this without prejudice, and though I had once on this subject to dip my pen in the ink of controversy, I am willing to accept with thankfulness the valuable instruction derived from the able critical examination of the plants that has thrown so much light on the comparative fossil vegetation of India and Australia, and this too in continuation of what long ago I believed to exist, the presence in the latter of true Triassic as well as Jurassic strata.*

Queensland and Western Australia.—Mr. Charles Moore (of Bath), F.G.S., enumerates 171 species of Secondary animal fossils from Queensland, all sent to him for description by myself; and sixty-two from Western Australia, of which twenty species are common to England and that Colony. (See Q. J. G. S, xxvi, 261.) (See *Appendix XIX.*)

In Mr. Dalrymple's "*Report of his Exploration on the North-east Coast of Queensland,*" (Brisbane, 1873, p. 20.), that enterprising observer states that the flat-topped ranges and mountains

---

* It will be seen from the following extract of a communication made by me to M. le Vicomte d'Archiac, 14 Nov., 1859, and which was published in the "*Bull. Géol. Soc. France,*" that I held opinions expressed as at the present respecting the position of *Glossopteris* in India mentioned by Dr. Oldham:—

"D'où l'on peut inférer au moins que ce genre ne caractérise pas seulement l'ère Jurassique. Il peut s'étendre au-dessous aussi bien qu' au-dessus, et, prenant ces faits en considération, on ne peut pas y voir un motif opposé à ce que j'ai dit si souvent, que la formation Carbonifère de la Nouvelle-Galles du Sud ne peut être partagée comme le propose M. M'Coy, et que, tandis qu'elle montre de nombreuses analogies avec celles de l'Europe, elle en diffère par l'existence à cette époque de genres qui ailleurs se montrent seulement dans la formation Jurassique.

"*Il reste donc à faire aujourd'hui une comparaison attentive de ces espèces de plantes douteuses de l'Inde, de l'Australie, de l'Angleterre, j'ajouterai de l'Afrique où les Glossopteris se rencontrent,* dit-on, dans les couches à Dicynodon de Blaun-Kopf." (Extrait d'une lettre de W. B. Clarke, à M. d'Archiac: Bull., xviii, p. 669.)

Dr. Feistmantel is now endeavouring to satisfy this desirable object in relation to India.

about the Endeavour River have "*red* sandstone escarpments," a feature that assimilates the formation somewhat to the "New Red" or Triassic.

The latter collection belongs chiefly to the Lower Oolites, Upper and Middle Lias; and the former embraces the Upper Oolites and Cretaceous formations. Mr. Brown, Government Geologist in Western Australia ("*Report of* 1873"), mentions Mesozoic beds in the Darling Range, and again on the South Coast, from Cape Rich to beyond Mount Barren and as far as Cape Esperance. Saliferous and reddish sandstones, &c., are the chief rocks. On his chart they and their detritus occupy seven degrees of latitude, and from one to three of longitude. But there is nothing defined as to fossiliferous evidence, except about Champion Bay. From Wizard Peak and Mount Fairfax I have received numerous fossils through the agency and kindness of the Hon. F. P. Barlee, F.R.G.S., Colonial Secretary, and the Rev. C. G. Nicholay, of Geraldton, who not only added to my collection, but supplied me with a personal survey of his neighbourhood on an enlarged scale, and with more minute details than Mr. Brown's chart exhibits. (See Q. J. G. S., xxiii, 7.)

South Australia and Tasmania.—There does not appear to be any fossiliferous evidence of Mesozoic formations in South Australia, where the rocks are chiefly Palæozoic, Metamorphic or transmuted, and Tertiary.

In Tasmania, there is, no doubt, about the same evidence as for New South Wales. Victorian geologists believe that the Coal of Jerusalem is Secondary. I was inclined to think that the neighbourhood of Green Ponds and Bagdad betrays a resemblance to some portions of the Wianamatta shales and sandstones of New South Wales. But the area there is far from extensive.

Mr. Gould, who surveyed considerable portions of the Colony, says nothing leading to the idea of any extensive Secondary areas; and whatever hold they may have on the mind of a geologist who has not carefully observed, must be due to preconceived notions as to the age of the Coal, some of which has of late established its Palæozoic character as unmistakeably as the seams of Anvil Creek, &c.

Coal has been reached on the Mersey under the Marine fossiliferous beds, as I always held it would be, in spite of vaticinations to the contrary.

New Caledonia.—Passing over to New Caledonia, the Secondary formations are represented by Triassic, Liassic, and Neocomian rocks or fossils.

On the 6th July, 1863, a paper by M. Eugene Deslongchamps was read before the Linnean Society of Normandy, on the Geology of Hugon Island, New Caledonia, in which mention is made of a Pecten and fish scale from Cape St. Vincent, on the S. S. W. Coast, collected by M. E. Deplanches. Millions of an Avicula (*Monotis*) allied to M. *salinaria* of Goldfuss, of which M. *Richmondiana*, of Zittel, is a variety, also occur. Astarte, Turbo *Jouani*, and one other; Spirifera *Caledonica* ; S. *Planchesi*; Scyphia *armata*—all these are Upper Triassic.

M. Garnier's fossils, examined by M. Fischer, were pronounced to be Monotis ; Halobia (an Austrian species) ; and Mytilus *problematicus* of the same formation.

The supposed Jurassic rocks contain Nucula near N. *Hammeri* (De fr.), a Littorina, a Cardium, and an Astarte resembling A. *Voltzii* (Goldf.) M. Fischer believes, however, that these are more likely to be Triassic also.

M. Munier-Chalmas names also as Jurassic Ostrea *sublamellosa*; Astarte (or Tæniodon) *præcursor*; Pellatia *Garnieri*; and Cardium *Caledonicum*.

A large *Pinna* seems to represent the Cretaceous rocks. A tolerably full account of the Geology of New Caledonia will be found in my "*Address to the R. Soc. N.S.W.*, 1875"; see also "*Note sur les Roches, &c.*," just published at Nouméa, by M. Ratte.

New Zealand.—New Zealand exhibits abundance of proofs that Secondary formations exist there, and not the least remarkable fact is, that Professor Hochstetter in 1859 discovered there the same Avicula *Richmondiana* as above, and Halobia *Lomelli*, Avicula *salinaria*, with Monotis, Spirigera, Spirifera, &c., belonging to the Triassic epoch.

In my paper "*On Recent Geological Discoveries*," I collected as much of this kind of information as I then could ; but since then the skill and labour of the Geological Survey of New Zealand, under the direction of Dr. Hector, have produced an abundant harvest of scientific details ; and to the able publications and reports from that authority I may refer those who are interested in the development of that most interesting group of islands. They will find there ample evidence as to the existence of Triassic, Jurassic, and Cretaceous, as well as of Palæozoic rocks. The Saurian discoveries of Mr. T. Hood Cockburn Hood, F.G.S. (see Q. J. G. S. xxvi, 1870, p. 409), also deserve commemoration ; nor must the labours and great discoveries of Dr. Haast, F.R.S., be unremembered.

So far as the Trias is concerned, Hochstetter's discoveries of the genera and species about Richmond have been rivalled by Captain Hutton, in Southland, Otago, who found in 1872, on the Moonlight Range, Monotis *Richmondiana* (Zitt.), and Halobia *Lomelli* (Wissm.) On the western slope of Hokanuis, and on the south side of the Wairaka Hills, he obtained the same species, with others, proving that the rocks are the same as the sandstones of Richmond, near Nelson, and also proving the Triassic age of the deposits. ("*Geology of Southland. Report of Explorations, Geol. Surv. N.Z.*," p. 104.)

Not very distant the same careful observer detected some of the same species as occur in Queensland in the Middle Jurassic formation, described by Mr. Moore, *e.g.*, Astarte *Wollumbillaensis*, with other genera and species that link in the South with the North Island (p. 105). These discoveries justify the inference that Triassic rocks *are* probably present also in New South Wales.

New Guinea.—It has long been known that Jurassic rocks exist at the northern end of New Guinea. But recently Signor d'Albertis brought to Sydney from the Fly River several fossils, among which Professor Liversidge noticed *Belemnites* and an *Ammonite* (of Liassic facies), &c. These I also saw—but I did not recognize those species which I have from Queensland.

CRETACEOUS.

When I first announced in 1860 the proof that Secondary fossils did exist in Australia, exhibited in Sydney, and forwarded to Sir Henry Barkly for Professor M'Coy's inspection, I especially mentioned the occurrence of Cretaceous species.[*] This was doubted, and the whole series classified as "*not higher*" than the "*lower part of the great Oolite.*" But in 1866, the Professor himself announced from another part of Queensland the occurrence of two *Inocerami*, and two *Ammonites*, from the Flinder's River district. He also announced an *Icthyosaurus*, a *Plesiosaurus*, and a *Belemnitella*, from lower Cretaceous strata of the same district.

Mr. Moore says, of the Wollumbilla fossils, "That they all belong to the *Upper Oolite* may with safety be inferred, but the Cretaceous beds have a claim to be considered," and he established the existence of the genus Crioceras, which was first reported by me.

In 1872, Mr. Daintree, F.G.S., read his Notes on Queensland, before the Geological Society, the Marine fossils illustrating which were (as before stated) described by Mr. Etheridge, F.R.S., L. & E., F.G.S., Palæontologist to the Geological Survey of Great Britain. The number of Oolitic species recorded is six, and of Cretaceous twenty-five.

The expedition of 1872, in the Cape York Peninsula, in which Mr. Norman Taylor, of the Victorian Survey, was Geologist, has

---

[*] See papers by the author "*On Recent Geol. Discoveries in Australasia*," (1861) pp. 27, 48, and "*On Marine Fossiliferous Secondary Formations in Australia*" (Q. J. G. S., xxiii, 8.)

added to the list of Secondary fossils in Queensland. These were sent to me for inspection by the Minister for Public Works in that Colony, and at his request forwarded to the Agent General in London. They have not yet been fully described.

A still further amount of Cretaceous fossils, forwarded by Mr. Hann, the leader of the Expedition of 1872, to Mr. Etheridge, and a large collection in my own cabinet, remain yet to be determined.

This is sufficient to show the extent of Mesozoic formations developed since 1860.

Mr. Daintree reckons the areas of the Cretaceous and Oolitic formations in Queensland at 200,000 square miles; the Carbonaceous (Mesozoic) at 10,000, and the Palæozoic Carboniferous at 14,000, whilst the Devonian and Upper Silurian occupy 40,000. The two younger, therefore, are nearly four times as extensive as the older.

After the "Norman Taylor" collection had gone to England, I received three or four specimens from the Table Mountain, between Hann's Camps 41 and 42 (*"Northern Expedition Report"*), and forwarded them to the Queensland Agent General in London for inspection by Palæontologists at Home. Mr. Etheridge, the Palæontologist of the Survey of Great Britain, considers the fossils in that conglomerate rock to be a species of *Hinnites* like *H. velatus* and an *Ostrea* like *O. Sowerbyi* and that they belong to the Oolitic series. The same conglomerate, as I learn by a more recent arrival, occurs on the high ranges between the Palmer and Cooktown, under the deposit which Mr. Daintree calls Desert sandstone. It is a coarse rock containing broken shells in a sandstone full of partly rounded pebbles. Mr. Etheridge also considers the Walsh River series to be of Lower Cretaceous forms. Some specimens of plants supposed to be Glossopteris were also forwarded by me to Europe, with the shelly rock. Mr. Carruthers's determination is, that they were not of that genus, but rather a form of Tæniopteris nearly allied to *Stangerites ensis* (Oldham and Morris in the Indian Survey Memoirs), which Schimper calls *Angiopteridensis*. Another specimen which I did not see in the great collection, but of which I had a drawing from Mr. Taylor, was considered by several geologists in Queensland, &c., to be Orthoceras, and, therefore, Palæozoic. Mr. Daintree says there were several specimens *like Orthoceras;* and so I think the one in question was, but I considered at the time that there was no Orthoceras present in the box, but a good many Belemnites, and I considered the sketch referred to was of the same genus.

I have since received the following statement—" There was no specimen of Orthoceras in the entire series."

I have also received a list of the genera of Walsh River fossils, in Mr. Etheridge's handwriting. It is as follows, making all of them Lower Cretaceous :—

Ammonites, allied to A. *Clypeiformis*.
Ammonites sp.
Crioceri.
Belemnites.
Myacites.
Byssoarca.
Solemya or Iridina.
Arca.
Panopæa.
Inoceramus.
Hinnites or Avicula.
Cytherea.
Cyprina.
Myoconcha.
Pecten.
Teredo or Teredina, in fossil wood.

An opinion has been adopted that the Mesozoic fossils from Queensland, both those described by Mr. Moore and those by Mr. Etheridge, were in mere drifted nodules. Mr. Taylor assures me that such is not the case with the latter, and I long ago gave a section of the beds at Wollumbilla, proving, as in the York Peninsula, that the nodular masses were derived from a soft shale, being in fact concretions. If they have been drifted they have not travelled far.

Mr. Taylor ("*Hann's Report*," p. 13) seems to have found the shelly deposit before mentioned on "a flat-topped Carboniferous range" (on 9 Sept., 1872) ; and by a report of April, 1875, from Cook Town, it appears that a fine seam of bituminous Coal has been discovered at the junction of Oaky Creek and the Endeavour River, 20 miles from Cook Town ; but from the determination of Mr. Carruthers, this Coal (confirming, however, Mr. Taylor's statement) is not of the Glossopteris age. The Coal of the latter series is not known to extend further north than 20° 35′ south.

In 1877 Professor Liversidge received from the Rev. G. Brown from New Britain and New Ireland (lat. 48° S. and long. 150° E.) some grotesque figures "cut by the natives out of a soft white pulverulent material," said to be thrown up by earthquake waves, and "having the appearance of plaster of Paris." It holds numerous remains of Foraminiferæ.

The account of it is given in an interesting paper read before the Roy. Soc. New South Wales, and published in their Journal 1877, vol. xi, pp. 85–91, "*On the Occurrence of Chalk in the New Britain Group.*" An analysis is given in comparison with English chalk, which it certainly resembles ; but a doubt may be expressed as to its being true chalk. Something like it, but less cretaceous, has been found in New Zealand, and I have found white calcareous fragments in the drifts of N. S. Wales resembling it. Professor Liversidge adds, that no true chalk has yet been found in Queensland or New Guinea, and I doubt whether it is older than Tertiary, probably such as the white beds of the Australian Bight or of Aldinga.

Mr. Brady, F.R.S., states that the Foraminiferæ are nearly all South Atlantic deep-sea species; there were other fossils also found during the "Challenger" Expedition.

§ 6. Tertiary Rocks.
Kainozoic of Duncan.

Throughout the whole of Eastern Australia, including New South Wales and Queensland, no Tertiary *Marine* deposits have been discovered. There are, however, in various places of New South Wales patches of *plant deposits* which, according to the frequent notices of geologists, may be referred to some period of the Tertiary epoch. A silicified sandstone or quartzite of this kind, full of impressions of ferns and leaves of trees, but not known to be now living, occurs at Jerrawa Creek not far from Yass. It is probably Miocene. On the summit of the Cordillera, near Nundle, about the Peel River Diggings, occurs a ferruginous bed full of leaves. On the Richmond River occurs a white magnesite, full of yellowish impressions of leaves. At Kewong, in the county of Gowan, there is a bluish deposit of fine aluminous matter with black impressions. From a depth of 60 feet in a shaft near Bungonia, a pale yellowish white deposit with similar impressions was brought up; and on the summit of a "made" hill, above Kiandra Gold Field, at a height of 4,000 feet above the sea, and in a region now partly covered with snow many months in the year, there is a deposit of black clay with such casts of leaves as occur in similar clay near Hyde in New Zealand.

In recent visits to various gold-fields in the Western districts, I have found plant-beds of somewhat similar kind either cut by the shafts or distributed in the wash-dirt below the alluvial deposits, underlying in some cases thick masses of basalt. Such occur at Gulgong; at Cargo; under Bald Hill at Hill End; and also at Blayney.

At Lucknow also occur deposits of branches and fragments of trees under the basalt, and on the Uralla Gold-field, and at Home Rule, on Cooyal Creek, lignite and woody matter of a similar kind were seen by me in the lowest deposit of the deepest shaft.

No botanist is willing to declare what is the exact age of such deposits; but some of the leaves are supposed to represent, among others, the foliage of *Fagus;* yet it was only in 1866 that a beech forest was discovered, by the Director of the Botanical

Gardens, growing on the Macleay River. On comparing the living leaves with the impressions in the various deposits mentioned, I can see no specific identity. This want of identity indicates, that however the plant may resemble living plants they cannot be of a recent period; and yet there are occasionally such close resemblances as to lead some good botanists to infer a recent period for some of them. These and some other Tertiary plants have been sent on at his request to Dr. Feistmantel, but too recently for learning his opinion.

The most remarkable instance I have examined is on the coast, about 42 miles north of Cape Howe, where, at a place called Chouta (between Tura and Boonda), a cliff about 100 feet high, formed of sand and white silicate of alumina, contains beds of lignite charged with sulphide of iron, and which are full of phytolites much allied to the living vegetation. From the clays, some of which are nearly kaolin, articles of pottery have been formed. It has been proved that, by distillation, a fair proportion of lubricating oil may be produced from the lignitiferous clay, and other products are expected to result from these deposits. The cliff is about 60 feet thick from the sea to the top of the clays, and borings below the sea-level have shown a still greater thickness.

These deposits lie between the horns of the little bay at Tura and Boonda, resting at one end on the highly undulating Palæozoic rocks, and at the other on a mass of porphyry. They were, formerly, no doubt, deposited in a depression among the slopes of the hills, but the wearing away of the coast has left a cliff of clay and sand instead of the original cliff of hard rocks. It is remarkable that at the south end, the rocks assume the character of a breccia of quartz cemented by siliceous matter (probably like a deposit mentioned by Mr. Gould as occurring in Tasmania) and in it analysis has detected the presence of gold, though some quartz veins at the north end contained none.

My impression at first was that the lignite is recent, but I place the deposits under the present head because it may be possible, notwithstanding the opinion of a botanical friend whose judgment is worthy of esteem, the plants are not recent. Baron Von Mueller, to whom I submitted them, hesitated to express an opinion. They are deposited in clays of various kinds, chiefly white. Some of the hardened clinker-like sands covering the clays remind me of the sands on the coast of Dorset, at Studland and Bournemouth. If this be really a Tertiary locality, it does not contradict the general assertion at the commencement of this section, for no shells of any kind have been detected in any part of these beds. Swampy and stunted plants still grow on the sands, which are very wet, and probably reproduce the phenomena beneath them, with the exception of the white clays which were in part derived from the decomposed felspathic matter of the porphyry. In various parts of Mancero there are lignite-like local thin deposits, but on analysis they have proved valueless.

By far the most interesting discovery that has been made in relation to the plant-beds was realized in the basaltic district of The Forest between Orange and Carcoar.

In the vicinity of the latter place are deposits of calcareous rock of the age of the Wellington Cave osseous breccia, also containing fragments of bone. I believe some specimens have been sent to Europe.

But these have not the same interest attached to them as the plant remains have.

The description of several new genera and species of these has been given in "*Observations on New Vegetable Fossils of the Auriferous Drifts*," by Baron F. Von Mueller, C.M.G., M.D., Ph.D., F.R.S., and L.S., Government Botanist, &c.; published by the "Mining Department" of Victoria, 1874. These have been discovered not only in The Forest, but also in Victoria, at Haddon, Nintingbool, Tanjil, and at Beechworth. They seem to belong to the later Pliocene formation, and to consist of plants allied to the present forest-belt of Eastern Australia. An abstract of the first account of them was read before the Geological Society on 22nd June, 1870, and afterwards copied from the Quarterly Journal (vol. xxvii) into the "*Geological Magazine*," 1870, p. 390.

They consist of the following species, viz.:—

Spondylostrobus	*Smythii*
Phymatocaryon	*Mackayi*
,,	*angulare*
Trematocaryon	*McLellani*
Rhytidotheca	*Lynchii*
,,	*pleioclinis*
Plesiocapparis	*prisca*
Celyphina	*McCoyi*
Odontocaryon	*Macgregorii*
Conchotheca	*rotundata*
,,	*turgida*
Penteune	*Clarkei*
,,	*brachyclinis*
,,	*trachyclinis*
Dieune	*pluriovulata*
Platycoila	*Sullivani*
Rhytidocaryon	*Wilkinsonii*

and, probably, some others.

This last species was discovered somewhere near Carcoar, in one of the gold-leads, in the beginning of March, 1875, on the 10th of which month I had the good fortune to re-discover it in the refuse from a shaft near Lumpy Swamp, in The Forest, between Orange and Carcoar. Baron Von Mueller having stated in his Report of 29th July, 1874, that we require to learn "what was the nature of their leaves and floral organs"; in order to search for these, I made a second journey to The Forest, having first explored it in 1872, and found, together with four specimens of *Rhytidocaryon Wilkinsonii* and a number of already described species, several *leaves* embedded in a ligneous clay in the refuse of a shaft, together with portions of the branches of some tree or trees. The tissue of the leaves was in some cases so thin that it peeled off on touching. The collection, which included a few other specimens of seeds and seed-vessels given to me by Mr. A. Montgomery, who lives in the neighbourhood, I sent on to the Baron, who said he would forward them to Professor Schimper, of Strasbourg, as he himself was unable at the time to undertake their examination. In a short time, therefore, we may expect to know more about these interesting plants.

The thickness of the rocks in The Forest and at Lumpy Swamp varies somewhat, but an example or two will show the character of the country over the gold-leads :—

Alluvium	10 feet.
Hard basalt	40 "
Decomposing basalt	40 "

Washdirt.

2. At Tigeroo shaft, near which I procured the seed-vessels:

Earth	10 feet.
Basalt	85 "
Peat and shale	10 "

Washdirt with seeds and leaves.

At Haddon, in Victoria, the fossil fruit was found in one shaft at the bottom of the following section, resting on Silurian slates. (See Lynch's plans, "*Vegetable Fossils of Victoria*.")

Black soil	1½ feet.
Red clay	4 "
Lumpy red and black clay	26 "
Clayey honeycombed rock, with hard cores, succeeded by zeolitic basalt	100 "
Do. decomposed at base	1½ "
Black clay	7 "
Drift gravel and sand (auriferous), Trees at the bottom	10 "
Auriferous wash dirt, (Fossil fruits)	6 "
	156

At Beechworth (El Dorado) occur wood and leaves in variably coloured clay above coarse drift, covering black clay with wood and leaves; and below this, two to eight feet of washdirt, holding fruits and woods, resting on granite. (From Mr. Arrowsmith's plan. *Id.*)

Professor M'Coy has enumerated in the list of Tertiary Victorian fossils between thirty and forty *Oligocene* species; thirty to fifty or more Miocene, together with many tropical types of Dicotyledonous plants; and from the auriferous drifts four Molluscs, six Marsupials, and a Dingo, with the wood and fruit of a Banksia and the foliage of Eucalyptus *obliqua*. These are partly *Pliocene* and partly *Post pliocene*. He has also figured and described a new Squalodon (*S. Wilkinsonii*) from the Cape Otway coast Miocene beds, and some species of Carcharodon from the Geelong district.

The occurrence of Banksia (four species) in the Tertiary formations of Hœring, in the Tyrol (see Clarke's "*Southern Gold Fields*," p. 173) and in Victoria, is a highly instructive fact as to the ancient vegetation of the world. The seed-vessels of plants deep below the surface of the auriferous drifts of Victoria and New South Wales were also mentioned by me in 1860, in the work alluded to above (p. 173).

In 1876 I collected a number of seed-vessels and leaves from the "leads" of Home Rule, and since then Mr. Wilkinson has made a considerable addition, from the auriferous deposits at Gulgong, to the species described, from the district of The Forest and Beneree.

Baron Von Mueller has described them as in the following list:—

Ochthodocaryon	*Wilkinsonii.*
Eisothecaryon	*semiseptatum.*
Illicites	*astrocarpa.*
Pentacoila	*Gulgongensis.*
Pleiacron	*elachocarpum.*
Acrocoila	*anodonta.*
*Phymatocaryon	*bivalve.*
*Plesiocapparis	*leptocelyphis.*
Spondylostrobus	*Smythii.*
Wilkinsonia	*bilaminata.*

The latter as well as other species of those genera marked * found also at The Forest. In addition to these another has been found.

Towards the north of the Cape York Peninsula the sandstones are barren of fossils, and about the Cape seem to have more the character of *Laterite*, resting on Porphyry.

Mr. Wilkinson, in his researches among the tin-mines of New England, recognized the drifts which in Victoria are considered Pliocene; and Mr. Norman Taylor and the late Professor

Thomson, in their paper "*On the occurrence of Diamond near Mudgee*" (*Trans. Roy. Soc. of N.S.W.*, 1870, p. 94) make mention of older and newer Pliocene drift. Whether there be any fossil evidence for the propriety of these terms I know not. That there are drifts of different parts of one epoch I believe; and, perhaps, the divisions are good, even if the designations be too refined. Dr. Duncan has advised us to postpone the Lyellian designations for the present. Having very recently visited almost every locality mentioned in that paper, and examined for myself much of the alluvia of the gold-fields in a large portion of the county of Phillip, I am prepared to testify to the extreme faithfulness of the description given by Messrs. Taylor and Thomson. My remark, therefore, about the term Pliocene is not to be taken as complaining of it, but as a justification for the introduction of some of the drifts in question under the present head. A distinction of time is however clearly marked in the character of the various deposits or in the difference of botanical remains.

Perhaps some of these deposits in the gold-fields, as well as some of the shelly conglomerates at the mouth of the Flinders, had better be considered as belonging to the next division of my subject; and though placed as Tertiary, I am not satisfied they are such, as no positive proof exists by unmistakable evidence that they are so.

In the far Western interior, beyond the Darling, shelly deposits of fine sandstone have been reached in well-making, and by the kindness of my friend Mr. Woore, C.C.L. of the Albert District, I have been put in possession of several good specimens, together with fossil wood, apparently not very ancient, which I believe to be Tertiary.

I have also from another contributor a very good specimen of a *Thalassina* resembling *T. Emerii*, from another part of "New Holland," which is said to have been found somewhere on the right bank of the Darling, not far from Mount Murchison. For the species alluded to, see the late Mr. Bell's paper in Q.J.G.S., 1, p. 93. Mr. B. received it from the late William S. Macleay, Esq., F.R.S., of Elizabeth Bay, Sydney.

Mr. Daintree has stated in his views respecting the Desert sandstone of his map, that it is a Kainozoic deposit, which once covered the greater part of Australia. In the places where it is in great force, in Northern Queensland, it overlies the Cretaceous rocks, and underlies lava beds. It contains fossil wood; and a *Tellina* which I sent to Mr. Daintree, from the neighbourhood of Leichhardt's crossing-place, on the Flinder's River, would, he says, if coming from the desert sandstone, show that that formation is not lacustrine. In various parts of New South Wales there are cappings of fine hardened sandstone which may have some relation to the strata referred to.

Mr. Daintree has, however, mistaken the locality he gives to the *Tellina*. He received a portion of a *Trilobite*, and not a *Tellina*, from Barkly's Table-land, and a cast of a whole one, which would give to that locality a Devonian character.

There is no doubt a fine waterworn drift over large areas of the auriferous and stanniferous regions and in the southern part of Maneero ; but in many cases the drift betrays its origin, as the result of the disintegration of conglomerates, and such I believe to be the origin of the drift seen by Professor Liversidge near Wallerawang. ("*Report on Iron Ore and Coal Deposits*," read before the Royal Society, 9 Dec., 1874.) He compares it with the diamond drift at Bingera, alluding to the "nodules of conglomerate" in each ; but this conglomerate may be found *in situ* in the Coal-bearing beds close at hand.

Many drifts have undoubtedly been dispersed, and re-agglomerated and again dispersed, from one age to another, and the fineness of the pebbles and their perfect attrition afford testimony as to their antiquity, though now called recent.

The outliers of the Tertiary deposits in N. W. Australia and what is called the "Northern Territory" (attached to South Australia) are little known beyond the coast, but there is probably a wide area between Cape Villaret on the North-west Coast and the watershed of the Victoria River in which Tertiary beds will be probably be found. The Rev. J. E. Tenison-Woods in 1864 [*] points out the Coburg Peninsula as Tertiary, and Port Essington was considered by Professor Jukes to have evidence of the same. Judging from collections in my own cabinets, there must be, however, a preponderance of far older formations. It is, nevertheless, also probable, from its auriferous conditions and the presence of granite and basalt, that there are Tertiary deposits in that portion of the interior, and of which the basalt may be the igneous representative. The Tertiary fossils of the South Coast of Australia, from near Cape Howe to Cape Leeuwin, have been partially known from the mention of them by several authors ; and those of South Australia and the Murray River have been more or less elaborately treated of by Sturt, Eyre, Angus, and with critical acumen by Woods, Busk, and Professor Tate of Adelaide. But somehow the great sections, nearly 600 feet thick, along the Australian Bight have yet to be catechised as to whether the Australian Tertiaries follow the laws which ruled the existence of these deposits in Europe, or whether the peculiar aberrations noticed by Mr. Woods in some of his valuable writings are or are not exceptions to those laws.

---

[*] "*North Australia: Its Physical Geography and Natural History.*" By Rev. J. E. Tenison-Woods, F.R.G.S., F.L.S., F.G.S., &c., *p.* 10.

New Zealand also contains a great number of Tertiary genera and species admirably detailed and arranged as belonging to the Upper Pliocene, Upper and Lower Miocene, and Upper Eocene, in a Catalogue by Captain F. W. Hutton, F.G.S. ("*Geological Survey, New Zealand,*" Wellington, 1873), of Tertiary Mollusca and Echinodermata, in the collection of the Colonial Museum.

The classification is based on the *percentage* of recent species, the proportions of which are 76, 34, 23, and 9 *per cent.*

With respect to the Australian Tertiaries, however, no one has done so much as the Rev. J. E. Tenison-Woods whose publications on the Victorian, Tasmanian, and South Australian strata are numerous and valuable. To enumerate them here would be unnecessary, as they will probably ere long be brought out by himself in a form available for the public benefit, and to the public appreciation of his long and persistent studies. Besides his numerous papers published elsewhere, Mr. Woods has contributed in 1877 to the Royal Society of New South Wales, no less than four papers showing great ability and very extensive knowledge of his subject. It appears from his researches, that there are peculiarities in the Australian beds, and that it would not be altogether safe in relation to Australian deposits to trust to European arrangements; nor does he think it clear that the Queensland cretaceous beds are altogether distinct from a commingling with Tertiaries. He has adopted also a view which must to a great extent be true, as to the sudden upheaval of portions of the Southern Coast of New Holland. As to the cliffs of the Australian Bight which have never yet been scientifically examined, there must have been at least 600 feet of elevation, but the fossiliferous beds appear to rest on granite, of which the slopes are abrupt, and which descend according to a statement made to me by the late Capt. Owen Stanley, R.N., F.R.S., to an enormous depth, of which mention is made in "*Geol. Magazine,*" vol. iii., pp. 503–551. Considering the depths sounded by the "Challenger," there is nothing remarkable in the idea that there may be depths within the assumed distance from the shore as great as any in the Atlantic or Pacific Oceans, or even greater, taking into account the general freedom from islands and reefs, regarding the granite as an evidence of upheaval, and its structure in vast nodular or spherical concretions. That such upheavals occupying even large areas along the Southern coast, are not inconsistent with subsidences of very great depth and extent on the East coast of Australia, offers no difficulty to those who regard such occurrences as the result of causes generally affecting the bottom of the ocean.

Professor Tate of Adelaide has already given a fresh impetus to the study of Australian Tertiary Geology, by his investigations respecting the Murray beds, and by the discovery among them of two species of fossils which have hitherto been held Cretaceous, an additional example of the manner in which certain genera ascend in time to overlying formations (see pp. 16, 31). The latest discovery of the existence of Tertiary Marine fossils is on the S.E. Coast of New Guinea. On the voyage of the "Chevert," the Hon. W. Macleay obtained a series of rocks and fossils, which I had the pleasure of seeing, and considered to be Tertiary. Since then they have been examined by Mr. C. S. Wilkinson, whose experience of the Victorian Tertiaries is so well known. He has determined ("*Proc. Linn. Soc. N.S.W.,*" vol. i., pp. 113-117) the following Lower Miocene shells, from Hall's Sound, most of which he recognizes as known in Victoria, and of which two have been described by Prof. McCoy ("*Prodrom. Dec. I*") :—

Voluta (*macroptera*)	Triton ?
„ (*anti-cingulata*)	Dolium ?
Ostrea.	Astarte.
Cytherea.	Corbula.
Crassatella ?	Læda.
Pecten.	Venus.
Turritella.	Cypræa.
Natica.	Echinodermata (2).

"*Notes on a Collection of Geological Specimens collected by W. Macleay, Esq., F.L.S., &c., from the Coasts of New Guinea, Cape York, and neighbouring Islands: By C. S. Wilkinson, F.G.S., Gov. Geologist.*"

The matrix of these fossils is described as exactly that of the Lower Miocene beds near Geelong and Cape Otway. At Katau, on the west side of the Bay of Papua, there are also fragments of shells in clay similar to those of Hall's Sound and Yule Island.

As described by Mr. Macleay, 11th Oct., 1875, Yule Island has a considerable inward dip from a horizontal face of cliff. The rock is calcareous, with corals, shells, echini, &c., bedded like the coral rag of Oxford. D'Albertis mentions basalt in the valleys, and coralline cappings on the hills, which reach a height above the sea-level of from 600 to 700 feet. In Victoria there is a similar arrangement—"Yellow and blue calcareous clays full of fossil shells, overlaid by thick beds of coralline limestone, consisting of an aggregate of comminuted fragments of shells and echinoderms."

Mr. Wilkinson regards the ferruginous capping of the porphyry of Cape York, which is but 90 miles distant from the Papuan coast, as Tertiary, and that the New Guinea beds may be yet found in the Cape York Peninsula. Of course, future researches may discover fresh deposits of Tertiary age, but so far as examination of the collections in my possession from Cape York, New Guinea, Brighton Cliffs, Flemington, &c., may serve as a guide, there is no proof of anything further than a resemblance in the colour and composition of the ferruginous sandstones of the Victorian localities to justify the supposition at present.

Dr. Rattray (Q.J.G.S., xxv. 297), in his "*Notes on the Geology of Cape York Peninsula*" (read 2nd May, 1869), says distinctly: "No fossils have been detected" in the sandstone "between the volcanic rock beneath and the superimposed Post-tertiary ironstone," in the bold cliffs of Albany Island and the opposite mainland. He mentions also that the Jardines, in their traverse of the Peninsula, found the same rock at various parts of their route; but he says also, that at the north end of Albany Island, where a boss of porphyry protrudes and displaces the overlying sandstone and ironstone, fine examples of chertified clay, ironstone, and quartzite may be seen at their point of contact" (p. 302.)

Now, Mr. Wilkinson gives a list of rock specimens as follows:

1. Quartz porphyry (Palæozoic) (?) from Cape York, found underlying beds of Tertiary (?) ferruginous sandstone.
2. Vesicular basalt and brecciated volcanic tufa (Upper Tertiary), from Darnley Island.
3. Small concretions of limonite, with polished looking surfaces, dredged up off the Coast of New Guinea.
4. Specimens of *Chalcedony and flint*, from Hall's Sound.
5. Oolite, limestone (Tertiary), very friable, from Bramble Bay.
6. Yellow calcareous (Tertiary), from Katau River.
7. Yellow and blue calcareous (Tertiary), from Yule Island and Hall's Sound.

Whether No. 4 has any relation to the "chertified clay ironstone" of Rattray I know not, but it is certain that there are many instances to be found in New Guinea of highly altered strata. No. 3 is also a common variety of iron ore in many places besides those indicated, *e.g.*, at New Harbour, 100 feet above the sea, where the nodules of iron have the exact kind of polish mentioned in No. 3, and are of considerable size. [Similar nodules, but of red species, occur also at Port Essington, on the opposite horn, so to speak, of the Gulf of Carpentaria.]

Although I do not go fully into particulars respecting evidence in my own possession concerning the Tertiary beds in the localities already mentioned, yet I may state that the calcareous rock of light colour occurs on various points between the coast and the Astrolabe Range, and, according to the data given me by officers of H.M.S. "Basilisk," near Redscar Head, at an elevation of 100 feet. I consider these beds to be Miocene also.

There are also junks of fossil-wood with thin veins of calcspar.

It may be well, in conclusion of this section, to allude to the facts pointed out in the previous parts, relating to the occurrence of genera and species in formations older than those in which they may usually occur.

In reference to such a contingency in Tertiary strata of Australia, the Rev. Mr. Woods in one of his papers seems to hesitate as to the passage into the Tertiaries from the Cretaceous, at the time of writing, he having seen no good grounds for the admission of such an occurrence. But since the date of that paper ["*History of Austr. Tert. Geol.*," read before Roy. Soc. Tas., 11th July, 1876], we find his admission ["*Journ. Roy. Soc. N. S. W.*," xi., 75, 1877] of two genera of generally considered Mesozoic age having been found in the acknowledged Middle Tertiary strata of Aldinga, in South Australia—species each of *Belemnites* and *Salenia*—discovered by Professor Tate [See Q.J.G.S. Feb. 7, 1877, xxx., p. 206.] He adds, that though Salenia was considered to be extinct, and a characteristic of Mesozoic form, "a living species was dredged up by the "Challenger.'"

Dr. Duncan, President of the Geological Society, remarked on the interest attached to the discovery of the Belemnite, which "added *another* to the curious examples of the survivors of older forms of life in Australia." As he expressed it, it was another of the Cretaceous forms "which had outlived the Cretaceous period. This and similar discoveries showed the impossibility of comparing Australian and English strata on purely Palæontological data." Other speakers confirmed the occurrence of such an apparent anomaly by facts from other localities.

Mr. Woods does not, however, think the doctrine of evolution can be sustained from Australian evidences, and has an explanation of his own, not revealed.

He says, further, that "during more than twenty years of researches in Australian Tertiary geology I have sought for any reasonable evidence in favour of evolution, or clue to its mode of operation, and have found none whatever. I must add, that Australian geology, whether reluctantly or not, must admit that she can urge nothing in favour of that theory being true, the true explanation of nature as we find it."

He concludes also, that "to assert that any part of the continent has been preserved as dry land since the Mesozoic period," would be a hasty conclusion, "and that the weight of evidence is against it." ("*Hist. Aust. Tert. Geol.*," op. cit., p. 25.)

§ 7. QUATERNARY FORMATION AND RECENT ACCUMULATIONS.

The Quaternary Fauna of Australia has been so long known by the patient and skilful researches of Professor Owen, that there is no need to do more than refer to his writings as the source of most of our knowledge respecting the strange animals that preceded the human epoch and perhaps extended into it. Huxley and others have also added to the general history of these creatures.*

The Diprotodon appears not to have been limited to any one portion of Eastern Australia, for its remains have been found in South Australia and Queensland as far north as the York Peninsula.

In many of the "gold-leads" also, fragments of bones are found. A section of one sample, at Wattle Flat, above the Turon River, is given in my paper on "*Fossil Bones*" (Q.J.G.S., xi., p. 405, 1855), and in "*Anniversary Address to Royal Society, N.S.W.*, 1873," p. 14.

One of the most recent discoveries of the extinct kangaroos is that of a portion of a skull of *Sthenurus minor*, from the district of the Castlereagh River, described by Professor Owen ("*Proceed. Zool. Soc., April* 17, 1877") as having relations to *Dorcopsis*

---

* An anecdote may be introduced here which may have some interest for visitors to the Australian Museum. In 1847, Mr. Turner sent to Sydney a box of bones from King's Creek, in Darling Downs, and Dr. Leichhardt, Mr. Wall (then Curator of the Museum), with myself examined them, and found there nearly the whole of the bones of the head, though in fragments only, besides other prominent portions of the Diprotodon skeleton, which had only been then partially known to Professor Owen, who had not at that time seen the *upper jaw*. So far, therefore, this individual was unique. With much trouble we put the bones together, and a cast was afterwards made of the skull, which is still in the Museum. A paper contributed by myself (dated 30th November, 1847), and afterwards re-published in the Appendix to my Report of 14th October, 1853 (" *On the Geology of the Condamine River*"), and some letters from the late W. S. Macleay, Esq., and Dr. Leichhardt, detailed the characters of the animal as far as they were then known, and the condition and other contents of Mr. Turner's collection. This would not deserve any mention here, but for the sake of introducing a curious event relating to the head of the Diprotodon alluded to. Mr. Turner sold his collection to the late Mr. Benjamin Boyd, who sent it to England. The ship was wrecked at Beachy Head, on the coast of Sussex, and the collection, forming part of the relics of the cargo which were sold, was taken to London, and Professor Owen bought it of the dealer who had become its owner, not knowing its history.

(Mueller). It was given to me by Mr. Lowe of Gooree. I forwarded it to Professor Owen, who deposited it in the British Museum as the type of the species.*

In many parts of the existing region, all over the surface, wherever the basalt rock is not denuded, also near Sydney, there are local deposits which might be called "till," were any Testacea found in them; and in the Interior there are widely spread accumulations of drift pebbles, which, as on the Hunter and Wollondilly, are rounded by attrition in their long journey from the mountains whence they have been derived. Sometimes, also, the breaking up of conglomerates has contributed to this drift.

On Peak Downs there are deep accumulations of drift, such as transmuted beds of the Carboniferous formation, igneous rocks such as porphyry and basalt, and fragments of the older Palæozoic formation. Many of these are encrusted with thin calcareous cement, which forms cups of clear calc-spar in hollows of a fine porphyritic grit—the same grit occurring on the Warrego, on the Ballandoon and Narran ridges, with transmuted quartzite, also in wells there and on the Darling near Fort Bourke, in which drift fine gold was detected by me to exist on the Downs, and has been again reported to me from the base of Rankin's Ranges on the Darling River, the furthest known Western auriferous locality in New South Wales.

In 1869 I reported the discovery of the femur of a bird, at the depth of 188 feet, in drift resting on granite, from a well in that part of Peak Downs (22° 40' S.) which lies between Lord's Table Mountain and the head of Theresa Creek, near the track from Clermont to Broad Sound. Compared with the bones of Dinornis in the Australian Museum, both the Curator of that institution, and myself came to the same conclusion as to its genus, and accordingly it was reported in the "*Geological Magazine*" as Dinornis. Professor Owen has, however, removed it into another genus *Dromornis*, considering it to have belonged to a Struthioid bird. If it was such, of course (especially after the deep soundings between Australia and New Zealand, established by H.M.S. "Challenger" in 1874), the speculations I indulged on a possible former connection between those countries as illustrated by such

---

* See "*Journ. Roy. Soc. N. S. W.* 1877," vol. xi., p. 209. With reference to this I have a communication from Professor Owen, dated 11th February, 1878, of which the following is an extract:—" I thank you for your timely appeal for the preservation of skulls and skeletons of the existing Marsupials prior to their *extinction*—that is but a question of time. Man is fated to that function, save in regard to such species, man inclusive, of which he can make any profitable use! It is an encouragement to study and to describe your fossils, to find 'Papers' so kindly commended as mine on *Sthenurus minor*. When shall we get a skull or jaw, or fragment of jaw with teeth, of your old 25-foot-long-lizard, *Megalania prisca*? It was contemporary with *Diprotodon*."

a discovery are worth little. But if it was a *Dromornis*, then it falls in with the relationship to a present bird, the Emu, just as the Kangaroos of this epoch are related in structure to the gigantic Marsupials of a past age.

[For *correspondence* connected with its first "identification," see "*Journ. Roy. Soc. N.S.W.*, 1877," xi., p. 45–49. See also a Memoir "*On Dinornis: containing a Restoration of the Skeleton of* DINORNIS MAXIMUS (*Owen*), *with an Appendix on Additional Evidence of the Genus* DROMORNIS *in Australia: By* PROF. OWEN, C.B., F.R.S., &c.," *Trans. Zool. Soc. Lon.* x., pt. iii., Oct. 1, 1877]

Mr. Thomas Cockburn Hood's discovery of crocodilian remains in New Zealand seems to establish in another way some possible connection long ago with distant regions, and crocodiles are yet in Queensland, the nearest probable land in the supposed insular or present fragmentary alliance with the former country.

The Northern coasts and islands would show also similar relations to New Guinea, and the only difference between the present conditions of such connections consists in the shallow seas of the present period in the latter, and the deep ocean between the points of direct communication in the other.

That the Pacific Ocean was formerly over wide areas now occupied as land has been a favourite view with many geographers; and although the Great Pacific Continent is rejected by others, yet there are not wanting additional proofs to sustain the decision, as to a great part of the ocean, as held by Fournier. (See *infra*.)

Africa and India, as well as Australia, New Zealand, and New Guinea, were probably in parts united. Not only do fossil plant remains add testimony to the probability, but the wingless birds, the reptiles, the vegetation of the present period, and the Marsupialia seem to connect the Northern regions, whilst, as Mr. Blanford shows in his interesting paper, "*On the Plant-bearing series of India or the former existence of an Indo-Oceanic Continent*" (see Q.J.G.S., xxxi., p. 510), a similar connection took place to the West.

It is not unsatisfactory, as to possible union of New Caledonia and New Holland, to find a similar view taken, upon grounds distinct from fossiliferous evidence or that of birds and reptiles. Under the head of "*Geographie Botanique*" in the "*Comptes Rendus des Sciences de l'Acad*^{ie.} *des Sciences*," tome 76, p. 77), there is a paper by M. Eug. Fournier, entitled, "*Notice sur la Dispersion Geographique des Fougeres de la Nouvelle Caledonic.*" The author gives a list of ferns special, as well as common to that group and to the islands of Polynesia and of the Pacific in general, &c., including New Holland, New Zealand, and Tasmania, in which latter group he finds 58 common to New Caledonia out of 289, showing that the latter is the head-quarters of those plants; and he reasons from this fact that New Caledonia was at some period connected with Australia by means of Norfolk Island and perhaps other submerged islands and with New Zealand and the Auckland Islands. "This hypothesis," he says, "will explain the simultaneous presence in lands at present under the influence of differing climates of species belonging to homogeneous groups, which could not by any causes have been transported by special currents, and which, living in the mountainous inner regions, are less exposed than littoral species to be carried away by exterior agents."

This hypothesis tallies completely with the possibility of the connection I presumed from the evidence of the supposed Dinornis,—which, however, is more strongly confirmed by Prof. Owen to be Dromornis, since he has examined, in addition to the *femur* from Queensland, a *tibia* from S. Australia, and the portion of a *pelvis* I sent him from N. S. Wales.

To the above considerations may be added, that Baron von Mueller having examined the plants brought from New Guinea by the Hon. W. Macleay, F.L.S., shows such resemblances with certain Australian species as to confirm M. Fournier's opinion respecting the former probable connection of the two great islands; this is properly referred to in Mr. Wilkinson's paper on the Geological Collections of the "Chevert Expedition," previously referred to (p. 97.)

The account of the plants by Baron von Mueller is to be found in three parts of a treatise entitled, "*Descriptive Notes on Papuan Plants: Melbourne.*" (Nov., 1875, to April, 1876.)

Remains of reptiles have also been found both in N. S. Wales and other parts of Australia, in Quaternary deposits, as for instance, *Megalania prisca* (Owen), a Lacertian allied to the Varans and Lace Lizards of Australia, which had, probably, a length of 25 feet; and in the great plains of the Interior bones of various gigantic marsupials, fishes, and reptiles are found bedded in black muddy trappean soil; and on Darling Downs, in Queensland, univalve and bivalve shells are found in some cases attached to the bones, or deposited over them in a regular series of layers at intervals of several feet; and of these shells some are yet living in the water-holes of the creeks. These facts are generally known, but it was not till recently that the osseous relics have been found in different creeks throughout the whole of the slopes and plains at the base of the Cordillera in Eastern Australia; in Victoria, in South Australia, and North Australia also. Of similar age are the accumulation of bones in caverns, as at Wellington; at Boree; near the head of the Colo River; at Yesseba, on the Macleay River; at the head of the Coodradigbee; not far from the head of the Bogan, and in other places.

A magnificent collection of the remains in the Wellington Caves has been made, at the instigation of Professor Owen, at the cost of the New South Wales Government, with the superintendence of the Trustees of the Australian Museum, by one of them, the late Professor Thomson, and by Mr. Gerard Krefft, F.L.S., C.M.L.S., &c., the late Curator of that Institution.

The Reports of these gentlemen, together with more than a thousand partly determined specimens, were forwarded to Professor Owen, who has expressed his acknowledgment of the value of this collection, "as regards novelty, instructiveness, and encouragement for the future," and as an "important element in working out the ancient history of the forms of animal life peculiar to Australia."

The Coodradigbee caverns will repay research hereafter. They have already furnished me with bones of birds, in which those of an Emu are prominent.

The latter fact chimes in with the alleged *Dromornis* of Queensland.

Professor M'Coy has named bones of a Dingo in a cavern near Mount Macedon. If it be really a dog of this period in Australia, it is another link between the Quaternary and Recent times. Vicomte d'Archiac, however, doubts its antiquity: "*Rien*," he says, "*ne prouve que ce chien n'ait pas été introduit par les premiers hommes qui ont peuplé le continent Australien.*" ("*Leçons sur la Faune Quaterniaire, Paris*," 1866, p. 271.)

An expedition to Howe's Island made known, in 1869, the existence of bones of birds and turtles embedded in the beach rock of the island. Afterwards, a collection of them was sent to me by Mr. Leggatt, of Fiji. I forwarded them to Professor Owen, who informed me that he was unable to determine to what they belonged owing to their imperfect state. They undoubtedly belong to some period near to the present, as the rock is a coral limestone common to the coast of the Pacific Islands; and that deposit also contains a Bulimus scarcely distinguishable from a living shell of the same genus off the Island, and eggs of Turtle also embedded as in Raine Island in the Barrier Reef. (See "*Trans., Roy. Soc. N.S.W.*," 1870, p. 37.)

Within the last few years, the drifts of the Cudgegong and Macquarie Rivers have been searched for diamonds, first reported in 1860 by myself as occurring in numbers in the latter river. Many thousand examples have been found; but they are chiefly small and of little value, though a few have been found of larger size and have been cut and polished.

A few have been brought to me from other localities in New South Wales, and some have been found in Victoria.

Mr. Norman Taylor examined the forms of the mineral as it occurs at Two Mile Flat, &c., and figured them with care. [See for collected references on Diamonds in Australia, in Professor Liversidge's paper, "*On Minerals of New South Wales*," *Trans. Roy. Soc., N. S. W.*, IX., p. 183, 1875; and W. B. Clarke's "*Ann. Address*," 1872.]

In other publications I have treated of them; and since then the Bingera Diamond Field has received careful attention from Professor Liversidge, who has described its condition accurately.

Those found since 1860 have fully justified the heading of my notice published that year ("*Southern Gold Fields*," p. 272),— "NEW SOUTH WALES A DIAMOND COUNTRY."

The Rev. Dr. Bleasdale, F.G.S., has published many valuable details of the Victorian gems.

Looking to the Colony of New South Wales, we find that in more than one instance the present river channels have deepened since the drift first began to crowd their banks. I have traced one of these drift streams, sometimes at great heights above the valleys, for more than 80 miles. In some places I have found upon the surface, as Strzelecki did in other parts, minerals (especially ores of copper, tin, and lead) which were at a great distance from their sources; and in two instances, that rare mineral, Molybdate of lead, of which no habitat has ever been yet found; and about three years ago a lump of Sulphuret of antimony, weighing three pounds, and exhibiting surface evidence of its being a drifted substance, was disinterred from the superficial ironstone gravel of an unfrequented place on one of the heights of the north shore of Port Jackson.

Some years since I reported on the occurrence of mercury in this Colony; but my expectation of the discovery of a lode of Cinnabar has been disappointed. The Cinnabar occurs on the Cudgegong in drift lumps and pebbles, and is probably the result of springs, as in California. In New Zealand and in the neighbourhood of the Clarke River, North Queensland, the same ore occurs in a similar way. About 1841 I received the first sample of quicksilver from the neighbourhood of the locality on Carwell Creek, on the Cudgegong, where the cinnabar is found. I proposed a full examination of that locality when I was in the neighbourhood in February, 1875; but the state of the weather was such as to preclude the possibility of doing so during my limited stay. But I was informed that the progress of the mine was satisfactory.

As connected with the drifts may be mentioned the occurrence of gems of all kinds in all the rivers where auriferous deposits occur, and subsequent years have only served to abundantly confirm my statement of 1860 as to the general distribution of them in the gold-bearing districts.

In examining the gold alluvia at a variety of shafts about Gulgong, Home Rule, and other places in the county of Phillip, I was struck by three prominent circumstances which have bearings upon the present and future of that region:

1. No shaft is, so far as I learned, deeper than 200 feet.
2. The gravels of the alluvia were composed of pebbles and fragments of rock common in the vicinity—derived from Carboniferous and underlying strata, with occasional fossils.
3. The quartz pebbles were in some cases perfectly rounded; in others the quartz was in fragmentary lumps, as if recently broken from reefs. These did not appear to occur together.

The conclusion I drew from the latter fact was that two periods of destruction and one of abrasion of underlying reefs had taken place at an early period of alluvial deposition. A fourth circumstance might be commented on. In the deposits of the shafts a multitude of well worn abraded lumps of jasper, silicified fossil wood, and semi-opal of various tints and chalcedonic interchanges, in some instances themselves decomposing so as to exhibit the fibres of the wood from which they had been formed by transmutation, arrested attention, and showed that either an older series of Carboniferous rocks had suffered such changes, or the beds of the series which now exhibits itself in various outliers had undergone the process. Mr. Lowe, of Gooree, has made a most extensive collection of these altered fragments, in which are many very beautiful specimens. It will probably never be rivalled, as he collected them from time to time as they were disinterred by the diggers. A great number also were coated with a shining transparent envelope of what I believe to be a deposit from silicified water. Elsewhere ("*Trans. Roy. Soc. N.S.W.*," 1870, p. 11) I have dwelt upon this; and it also attracted the attention of Professor Thomson and Mr. Norman Taylor. These deposits are frequently covered by a great thickness of basalt, upon which frequently lies a more recent drift partly derived from older drifts. The colours of the alluvia, now long exposed, rival in some degree those poikilitic hues which distinguish the west end of the Isle of Wight.

A drift of local kind also occurs over large areas in Maneero in the neighbourhood of the auriferous strata, as also in New England over the country of the tin-mines, which exhibits the same sort of alluvia as the gold-fields, and in which also gold occurs. In 1851-3, when I discovered the first tin in the Colony, it was generally in association with gold and gems. Messrs. Ulrich, Wilkinson, and Liversidge have since that time made local explorations both in the alluvia and in the beds from which they have been derived. There are deposits of opals besides those in the gold-drifts; and on Lawson's Creek, a feeder of the Cudgegong, agate breccias and opals occur. Opaline veins also occur in the basin of the Abercrombie River; in that of the Barcoo, in Queensland; and about 25 miles S.E. of Cudgellegong Lake, on the Lachlan River.

At the mouths of the Richmond and Clarence Rivers gold is found distributed in the sands and covering pebbles of the sea beach; a similar distribution is found in the sands of Shell Harbour, Illawarra (where the accumulations above-named occur), and some gold was extracted. Other spots give similar indications; and one specimen of gold was brought up from the sea bottom by the sounding operations of H.M.S. "Herald," off Port Macquarie. Gilded pebbles also occur on the West coast of New Zealand.

Numerous instances have also been recorded of gold having been found in the gizzards of wild fowl and of domestic poultry, in various parts of the Colony, confirming, with the abovementioned facts, the almost universal distribution of the precious metal in river-drifts and superficial deposits. Some of the above-named examples of gold collected by birds were exhibited by me at Sydney and in Paris in 1855, and are still in my possession.

All along the coast, from Torres's Strait to Bass's Strait drift pumice may be found wherever there is a lodgment, generally in the north corner of the little shore bays. That this has gone on for ages is apparent, as in one part of the coast north of Wollongong there is an accumulation of water-worn pumice some distance from the shore, and beyond the reach of the present waves. It is supposed to come in during easterly gales, from the volcanic islands to the north-east. In 1841 this fact, and all the evidence then collected in relation to such drift and "atmospheric deposits of dust and ashes," were published in a paper I forwarded to the "*Tasmanian Journal*," of which D'Archiac ("*Prog. de la Géol.*") was pleased to say it contained all that was known on the subject.

Subsequently received facts have only confirmed what was then stated.

Along the coast of New South Wales are found ranges of dunes, with a variety of shells, some of them rare, others common, as on Port Hacking and Cronulla Beach; along the shores of Botany Bay; on the great flat between the Hunter and Port Stephens, and along the Macleay River, which now passes for many miles through the shelly accumulations; and about Moreton Bay, and in more northern coast openings, shells and Marine refuse form deep deposits, from which, as in Illawarra and Broken Bay, a considerable profit is obtained by dredgers and shell-collectors, for the production of lime.

Respecting recent species very little is actually known of many of them, comparatively speaking, in any of the Colonies.

Of course there are scattered notices and collections in Museums. These, however, have not yet become historical. The Rev. J. E. T.-Woods, is the only writer who has taken up the subject systematically in Tasmania, and we are indebted to him for some valuable lists, of which may be mentioned "*Description of New Tasmanian Shells* (*Proc. R. S. Tasm.*, 1875 and 1876)," and "*Census, with brief descriptions of Marine Shells of Tasmania and adjacent Islands* (*Proc. R. S. Tasm.*, read 13 Mar, 1877)." This, indeed, comprises long lines of the coasts of East, South-east, and West Australia, embracing upwards of 550 distinct species, besides those of the Tasmanian coasts and Bass's Strait, collected from various sources and especially from the compiler's personal researches.

The late Mr. Strange was a collector of Port Jackson Shells; but what became of his complete collection I do not know. Some are, probably, in the Australian Museum, Sydney. Mr. Brazier, C.M.Z.S., the Rev. R. L. King, B.A., Mr. Hargraves, Junior, and others, have also contributed many species; and on the whole, the catalogue must be an extensive one.

Raised beaches also occur at various heights on rocky projections of the coast, indicating elevation of the land, of which there is distinct evidence in the recent period, not only in Moreton Bay, but near Sydney, and thence to Bass's Strait; also on both sides of that Strait, and as far as Adelaide and King George's Sound. Mr. Selwyn gives data for assuming the elevation of the land to have reached occasionally 4,000 feet in Victoria, but he has no evidence of Tertiary Marine fossils above 600 feet. Unfortunately, on the Eastern coast, having no Marine Tertiaries, we have to found our deductions, as respects New South Wales, on less secure data. Yet we have here evidence of another kind, and pot-holed surfaces of considerable extent have been found by me at various heights from 300 to nearly 3,000 feet.

In a brief memoir like the present it is impossible to quote all the authorities, nor has time allowed a more satisfactory digest or a wider range of statements. What has been thus collected is brought together in the design of giving a concise summary of the general Geology of the Colony, omitting, on account of its perplexity, all specific reference to the igneous rocks traversing, covering, transmuting, or supporting the Sedimentary deposits.

In *this* Edition many new facts have been introduced with the view of bringing on the discoveries that have been made from time to time to the present period, when a new system of geological inquiry has been just instituted in this Colony.

If private independent travel and research have not been able to accomplish more than this abstract discloses, it may be hoped that now the Government has commenced the work from its own resources, pecuniary and official, more will be accomplished than has hitherto been done to work out the intricacies of Australian geology, to accomplish which in minute and thorough detail will probably require the united exertions of many a worker in the field and the cabinet to the end of the next century at least. In the preceding pages it has been my lot to mention many of my own discoveries; but it has not been with any desire to overrate my endeavours or exertions; and some I have altogether omitted.

In the *first* Edition of this paper mention was merely made of the Cape York Peninsula, where ferruginous deposits occur on the lower slopes and bases of porphyry hills. I may repeat here what was added in the *second* Edition. Those deposits were examined at the Mint, and no gold was detected; but on a recent comparison of their lithological character with that of Tertiary beds from Flemington (in Victoria), I believe them to be, if not Tertiary, of similar origin to the *Laterite* of India, and of the Islands in the intermediate sea.

Dr. Rattray, of H.M.S. "Salamander," who furnished me with a map, and a collection to illustrate it, from the neighbourhood of Cape York, and whose paper was read by me, in his absence, before the Royal Society of New South Wales, more recently published his views *in extenso* before the Geological Society of London. He therein attributes to me an opinion that the thick sandstones of the Peninsula are of the age of the Hawkesbury rocks of New South Wales.

I do not remember that I have expressed any opinion on this sandstone; what was submitted to me was considered by me far younger. That such sandstone, and even older deposits between Cape York and the Gilbert River, may exist in the interior of the Peninsula, is far from improbable. The data at present are insufficient for further comment. It may belong to the Desert sandstone of Daintree.

But this inference may be permitted, that as Cape York is so short a distance from the gold-bearing deposits of New Guinea, and as, as is now proved, all the rivers running to the Gulf of Carpentaria from the Mitchell to the Nicholson inclusive rise in auriferous ranges, gold will probably be found in some parts of the country along the back-bone of the Peninsula; and although my past examination of the rocks in the Louisiade Archipelago has not proved gold to exist there, yet I agree with Mr. Daintree in his last Report to the Queensland Government, that the strike of the older formations justifies the belief that the Archipelago, and, I may add, other portions of the lands insulated in that part of the Pacific, will eventually furnish their quota of the precious metal.

Several collections of New Guinea rocks have been sent to me; but although it was asserted strenuously that gold was found in them, in the district visited by H.M.S. "Basilisk," I have not been able to recognize the existence of any auriferous matrix, though it is well known that alluvial gold was discovered during the visit of H.M.S. "Rattlesnake" on the coast at the other side of the Island. I find, however, that nodules of excellent hæmatite occur at New Harbour, about 100 feet above the sea. We have had satisfactory additions to our knowledge of that great Island from the results of the Expedition so nobly undertaken by Mr. Macleay.

In 1870 I added a remark or two about the discovery of a living Ceratodus in the waters of Queensland in the preceding year, the only previous known existence of the genus being the *teeth* found in Triassic European rocks to which that name was given.

This was an interesting addition to the living Trigonia Cestracion, Terebratula, &c., of Australia, which connect the present period with the forms of life once held to be extinct.

Inquiries respecting this curious fish have resulted in the discovery of other species than that first found (Ceratodus *Forsteri*), and what is more extraordinary, fossilized teeth, of which I was shown examples by Professor Wyville Thomson, who found them in an excursion purposely undertaken in search of the fish during the stay of H.M.S. "Challenger" in Port Jackson.

Since the first description of the fish by Mr. Krefft, Dr. Günther, F.R.S., has published a valuable "*Description of Ceratodus, a genus of Ganoid Fishes recently discovered in rivers of Queensland, Australia,*" in the "*Phil. Transactions*" (part II., 1871). The result is, that both Agassiz and Pander had, from teeth found in the Lias and Trias of Europe, come to conclusions which the living Ceratodus fully justifies. Dr. Oldham also had reported Ceratodus teeth from Maledi, south of Nagpur, in India. Australia in this instance precedes India. The fish turns out to be allied to Lepidosiren, and its habits are amphibian as it feeds on grasses and weeds in fresh water.

Dr. Günther goes into a most elaborate and minute examination of the anatomy of all parts of the fish, and a comparison with other fishes of the same and different types. He sums up thus—"The Dipnoous type is represented in the Devonian and Carboniferous epochs by several genera (*Dipterus, Cheirodus, Conchodus, Phaneropleuron*); it is then lost, down to the Trias and Lias, where the scanty remains of a distinct genus, *Ceratodus*, testify to its presence; no further trace of it has been found until the present period, where it re-appears in three genera, one of which is identical with that of the Mesozoic era. Now, at present, scarcely any zoologist will deny that there must have been a continuity of the Dipnoous type; and it is only a proof of the incompleteness of the palæontological record, that we have to derive all our information regarding it from only three so very distinct periods of existence. The *Dipnoi* offer the most remarkable example of persistence of organization, not in fishes only, but in vertebrates. On a former occasion I have shown that numerous recent species of fishes have survived from the period of the geological changes which resulted in the separation of the Atlantic and Pacific by the Central American Isthmus. In Ceratodus we have now found a *genus* which, as far as evidence goes, persisted unchanged from the Mesozoic era; and in the *Sirenidæ*, a *family* the nearest ally of which lived in the Palæozoic epochs."

This is a most valuable link in the connection of the old geologic periods with the present era, and a fit conclusion for the account above given, however unworthy that account may be, of Quaternary and Recent accumulations.

No general notice in this Memoir has been taken by me of igneous rocks; but it may be suitable to state that there is, in all the various Sedimentary formations noticed, distinct evidence of the presence of igneous action (*hydro-igneous* rather), and their transmutation through such and allied agencies has left an impress upon all the rocks more or less concerned. Such references will be left to another occasion.

No particular or special reference could enter into the object for which this Memoir is written; but it is to be understood that, though all the rocks have undergone a transmutation, this does not constitute what geologists have understood by "Metamorphic" system, of which, as before said, New South Wales, at least, shows little or no visible trace.

In order to explain the position of Glossopteris in the Palæozoic Marine deposits, I have appended two vertical sections, one, by myself, previously published in the "*Transactions of the Royal Society of Victoria*," 1861, illustrating the Coal-seams at Stony Creek; and the other showing the deposits at Greta, near Anvil Creek, which has been reduced from one on a larger scale kindly supplied to me by Mr. James Fletcher, Colliery Viewer, to whom I am also indebted for a collection of strata, the characteristics of which I have given after careful examination of them and of other specimens collected by myself on former occasions. The latter section illustrates a wide area on that part of the Hunter River. No. 2 is about 10 miles west of No. 1.

I have also appended two sections, one from Mount Victoria and the other from Burragorang, as well as a map showing portion of the Wianamatta Basin—which were made to illustrate my paper on "Oil-bearing Deposits" cited at p. 68, but which were not then published.

APPENDICES.

I. Collection made by Sir T. L. Mitchell, 1831-1836.
II. New South Wales Fossils, collected by Dana, 1839-40.
III. List of numbers of specimens forwarded by Rev. W. B. Clarke, in 1844, to Cambridge, collected during 1839-44.
IV. List of Fossils, recorded by M. de Verneuil, 1840.
V. Leichhardt's List, 1842-3.
VI. Wollongong Fossils, recorded by J. Beete Jukes, 1845.
VII. Carboniferous Flora, Upper Coal-beds overlying Palæozoic Marine beds. List by Morris, 1845; collection made by Strzelecki.
VIII. "A." Carboniferous Marine Fossils. List by Morris, 1845; collection by Strzelecki.
IX. M'Coy's List, 1847: "Coal Measure Plants," collected by W. B. Clarke.
X. M'Coy's List, 1847: "Wianamatta Plants," collected by W. B. Clarke.
XI. "B." M'Coy's List, 1847: "Carboniferous Marine Fossils."
XII. Stutchbury's "Devonian Fossils," 1851-3.
XIII. Plantæ. (1.) "Upper Silurian." (2.) "Devonian." (3.) Between "Upper Devonian" and "Lower Carboniferous." (4.) "Carboniferous."
XIV. De Koninck's "Upper Silurian Marine Species," N.S.W.
XV. De Koninck's "Devonian Species," N.S.W.
XVI. "C." De Koninck's "Carboniferous Species," N.S.W.
XVII. Lonsdale's List, N.S.W., of *Zoantharia*, 1858.

Extract from letter, 12 July, 1858, from W. Lonsdale to W. B. Clarke.

Salter's Notes on same, "Upper Silurian" and "Devonian" Species, borrowed from Woodwardian Museum, Cambridge.

Remarks on the preceding Lists, by W. B. Clarke.

Extracts from letters by J. W. Salter, of 9 May, 1856, and 28 November, 1858, to W. B. Clarke.

XVIII. Schemes of arrangement, by different authors, of the Palæozoic Fossils of New South Wales Sedimentary Formations.
XIX. Mesozoic Marine Fossils: Lists by Chas. Moore, F.G.S.—Western Australia and Queensland.
XX. Correlation of Australian Fossils, and Systematic Table: By Ottakar Feistmantel, M.D.

APPENDIX I.

COLLECTIONS made by Sir T. L. Mitchell, Sur. Gen., during his Expeditions of 1831 and 1836, determined for him by the late W. Lonsdale, Esq., F.G.S., Curator of the Geol. Soc., London (see vol. 1, pp. 14-16; and "Report" to Government, of 16 Octr., 1851) :—

	Genus.	Species.	Locality.
Carboniferous.	Plant impressions......	....	Broken Back; Hunter River.
	Glossopteris	Browniana ...	
	Fossil wood	....	Kingdon Ponds; Harpur's Hill; Minamurra R.
	Lepidodendron ? drifted	.... *	Road between Windsor and Parramatta.
	Lithostrotion	....	Ridge below Perimbungay.
	Crinoidal stems	....	S. of Perimbungay.
	Spirifer	glaber	Harpur's Hill.
	,,	sp.	Mount Wingen.
	Isocardia ?.............	sp.	Harpur's Hill.
	Litorina...............	filosa	Perimbungay.
	Megadesmus .... = *Pachydomus*.........	antiquatus...... cuneatus ...... globosus ...... lævis	Harpur's Hill
	,,		
	,,		
	,,		
	Terebra ?	....	near Mulucrindie.
	Trochus................	oculus	Harpur's Hill; Williams R.
	"Shells"	....	Bunuemir Ck. Wollondilly.
Devonian ?	Lepidodendron.........	....	Bed of Peel R. at Wallamoul.
	,, ...	....	near Honeysuckle Hill.
	Favosites	Gothlandica ...	Limestone Plains.
	,,	alveolaris	Shoalhaven gullies.
	,,	sevl. other sp.	Shelley's Cave, Argyle.
	Stromatopora	concentrica ...	Limestone Pls.
	Heliopora	pyriformis......	Coodradigbee R.
	Crinoidal stems	....	Do. & Limestone Pls.

* In sandstone. Similar occurrence on the Warragamba, above junction with Nepean.—W.B.C.

APPENDIX II.

Fossils of New South Wales, described by Professor James D. Dana. Collected by him, 1839–1840.

(See *Appendix I.*—"*United States Exploring Expedition.*")

Genus.	Species.	Authority.	Locality.	Remarks.
Urosthenes (Dana)	Australis	Dana	B Coal Pit, Newcastle	Taken out by Mr. Jas. Steel.
			1. PISCES.	
			2. MOLLUSCA. (*a*) BRACHIOPODA.	
Terebratula	amygdala	Dana	Black Head; Illawarra	
,,	elongata		Illawarra.	
,, (?)	(?)		ibid.	
,, (?)	sp.		Glendon.	
Spirifer	glaber (subradiata of Sow.)		Black Head; Harpur's Hill	Abundant, Mt. Wellington, Tasmania (Morris); Darlington, N.S.W. (M'Coy). Resembles *S. Hawkinsii* (Devonian).
,,	Darwinii	Morr.	Glendon	Near *Darwinii*.
,,	duodecimcostatus	M'Coy	Muree; Wollongong	Near *acuticostatus* (De Koninck).
,,	sp.		Black Head; Glendon.	One spec. resembles *avicula* (Sow.); E. H. Neck and Korinda (M'Coy).
,,	vespertilio	Sow.	Black Head; Eagle Hawk's Neck, Tas. (Morr.).	Referred by Morris to *vespertilio*.
Siphonotreta (?)	phalaena	Dana	Illawarra	
Lingula	curta	Dana	Glendon.	
Productus	ovata	Dana	Black Head.	Near *lata* (Murchison).
,,	fragilis	Dana	Wollongong Point.	
,,	brachytherus	Sow.	ibid.	
			MOLLUSCA (*b*) ACEPHALA.	
Solecurtus (?)	ellipticus	Dana	Wollongong Point.	Doubtful.
,, (Psammobia ?)	planulatus	Dana	Harpur's Hill	Doubtful.
Pholadomya (Platymya)	undata	Dana	Wollongong Point.	Doubtful; perhaps near Maconia.
,, (Homomya)	Glendonensis	Dana	Glendon.	
,,	audax	Dana	Wollongong Point.	Doubtful; may belong to Allorisma (King).
,,	curvata (?) (Morr.)	Dana	ibid.	
Astarte	gemma	Dana	ibid.	
Astartila (Dana)	intrepida	Dana	ibid.	
,,	cyprina	Dana	ibid.	Pachydomus *ovalis* (?) (M'Coy).
,,	cytherea	Dana	ibid.	Pachydomus (?) *pusillus* (?) (M'Coy).

APPENDIX II—continued.

MOLLUSCA (b) ACEPHALA—continued.

Genus.	Species.	Authority.	Locality.	Remarks.
Astartila (Dana)	polita	Dana	Black Head.	
"	cyclas	Dana	Wollongong Point.	
"	transversa	Dana	Wollongong Point.	
(?)	corpulenta	Dana	Illawarra.	
Cardinia (?)	recta	Dana	ibid.	Provisionally assigned.
(?)	cuneata	Dana	Wollongong Point	C. exilis (M'Coy).
(?)	costata (Mor.)	Dana		Orthonota (?) costata (Morr.)
Pachydomus	cuneatus (Sow.).	Morr.	Harpur's Hill	Megadesmus (Sow.)
"	antiquatus (Sow.)	Morr.	ibid	Id.
"	laevis (Sow.)	Morr.	ibid	Id.
Mœonia (Dana)	elongata	Dana	Black Head	Sub-genera Pyramia = Notomya (M'Coy) and Cleobis.
"	valida	Dana	ibid.	
"	axinia	Dana	Illawarra	= Cypricardia (?) sinuosa; resembles Pachyd. carinatus (Morr.)
(?)	carinata (Morr.)	Dana	Wollongong	Pachydomus carinatus (Morr.); Cypricardia rugulosa (Amer. Exp.)
"	fragilis	Dana	Glendon.	
"	myiformis	Dana	Wollongong Point.	= Pyramus myif. (Amer. Exp.)
"	elliptica	Dana	Harpur's Hill; Wollongong	= Pyramus ellipt. (Amer. Exp.)
"	gigas (M'Coy)	Dana	Illawarra	Pachydomus g. (M'Coy)
"	grandis	Dana	Wollongong Point; Flinders Island (1858)	= Cleobis g. (Amer. Exp.); Pachydomus globosus (?) (Morr. and M'Coy).
"	gracilis	Dana	Wollongong Point	Near gracilis = Cleobis gracilis. (Amer. Exp.)
(?)	recta	Dana	ibid	= Cleobis recta. (Amer. Exp.)
Nucula	abrupta	Dana	ibid.	
"	concinna	Dana	Harpur's Hill.	
"	Glendonensis	Dana	Glendon.	
Eurydesma (Morr.)	elliptica	Dana	Harpur's Hill	Isocardia (Sow.) near Avicula.
"	globosa	Dana	Illawarra	Pachydomus sec. (M'Coy).
"	sacculus (M'Coy)	Dana	Harpur's Hill	Isocardia (?) (Sow.)
Cardium	cordata	Morr.	ibid	Pleurorhyncus Aus. (M'Coy); referred by him to Wollongong, but locality suspected by Dana.
"	Australe (M'Coy)	Dana	Glendon	Referred doubtfully to this genus.
Cypricardia	ferox	Dana	Wollongong Point	Modiolopsis acut. (Hall)
"	acutifrons	Dana	Illawarra	

APPENDIX II—continued.

Genus.	Species.	Authority.	Locality.	Remarks.
MOLLUSCA (b) ACEPHALA—continued.				
Cypricardia	imbricata	Dana	Harpur's Hill	Modiolopsis imb. (Hall).
"	arcoides	Dana	ibid	Modiolopsis arc. (Hall).
"	prærupta	Dana	Illawarra	Modiolopsis præ. near faba (Hall).
"	siliqua	Dana	ibid	Modiolopsis siliq. (Hall); resembles Mytilus (Modiola) Teplofi (Vern.)
"	simplex	Dana	Glendon.	Modiolopsis simp. (Hall).
(Avicula ?)	Veneris	Dana	Wollongong.	
Avicula	Volgensis (?)	Vern.	Illawarra.	
Pterinea	macroptera	Morr.	Harpur's Hill	
Pecten	comptus	Dana	Illawarra.	P. sub-5-lineatus (?) (M'Coy).
"	leminiscellus	Dana	Illawarra.	
"	tenuicollis	Dana	Harpur's Hill.	
"	mitis	Dana	Glendon.	
"	Illawarrensis	Morr.	Harpur's Hill	Morris gives Illawarra, which is doubted by Dana. M'C. cites Wollongong, but (? olive-coloured rock of Harpur's Hill, Dana).
"	squamuliferus (?)	Morr.	ibid	
" (?)				
MOLLUSCA (c) GASTEROPODA.				
Pileopsis	tenella	Dana	Harpur's Hill	Patella ten. (Amer. Exp.)
"	alta	Dana	ibid	From Rev. Mr. Wilton's collection.
Pleurotomaria	Morrisiana	M'Coy	Black Head ; Harpur's Hill	Pleuro. triflata (Amer. Exp.)
"	nuda	Dana	Illawarra	
Platyschisma	Strzeleckiana	Morr.	Harpur's Hill.	
"	oculus	Morr.	Illawarra.	
"	rotundatum	Morr.	Harpur's Hill.	
"	depressum	Dana	Illawarra	Doubtful.
Natica (?)	undulatus	Dana	Harpur's Hill.	
Bellerophon	strictus	Dana	Wollongong.	
"	micromphalus	Morr.	Illawarra	Murce, cited by M'Coy.
MOLLUSCA (d) CEPHALOPODA.				
Theca	lanceolata	Morr.	Black Head.	
Conularia	inornata	Dana	Glendon	Near C. irregularis (De Kon.)
"	lævigata	Morr.	Harpur's Hill	Illawarra (Strzel.), but doubtful according to Dana.
"	tenuistriata (?)	M'Coy	Harpur's Hill	Murce, cited by M'Coy.

APPENDIX II—continued.

Genus.	Species.	Authority.	Locality.	Remarks.
\multicolumn{5}{c}{**3. RADIATA.**}				
Fenestella	internata	Lons.	Glendon.	
,, (?)	sp.	Dana	ibid	Near *internata*.
,,	media	Lons.	ibid.	
,,	ampla	Lons.	ibid.	
,,	fossula	Lons.	ibid.	
,,	gracilis	Dana	ibid	
Chaetetes	crinita (Lons.)	Dana	Wollongong Point.	Near *formosa* (M'Coy). Stenopora *crin.* (Lons.)
,,	Tasmaniensis (Lons.)	Dana	Harpur's Hill.	(Cited from Mount Wellington, &c., Tasmania, by Lonsdale). *Tas.* (Lons.)
,,	ovata (Lons.)	Dana	Harpur's Hill.	Stenopora *ovata* (Lons.)
,,	gracilis	Dana	Wollongong Point; Black Head.	
Hemitrypa (?)	...	...	Glendon; Wollongong.	Doubtful.
Enerinital	reuniuis	Dana	Hunter River.	Occurs with Fenestellæ, at Glendon
Pentadia	coronn	Dana		
\multicolumn{5}{c}{**4. PLANTÆ.**}				
Coniferous wood.				
Fruit scales.	spatulata	Dann	Illawarra.	
Noeggerathia (Göpp)	media	Dana	Newcastle	(?) *cuspulia* (Göpp).
,,	elongata (Morr)	Dana	ibid	*Zeugophyllites elong.* (Morr).
Calamites (?) (Göpp.)	lobifolia	Morr.	Newcastle.	
Sphenopteris	Browniana	Brgt.	Newcastle; Illawarra.	"Constituting 9-10ths or perhaps 99-100ths of all the fossil leaves in these districts."
Glossopteris	ampla	Dana	ibid.	
,,	reticulum	Dana	Newcastle.	Reticulation like *ampla*.
,,	elongata	Dana	ibid.	
,,	cordata	Dana	Illawarra	
,, (?)	linearis	M'Coy	ibid.	
Phyllotheca	Australis	Dana	Newcastle	From Mr. Wilton's collection.
Clasteria (Dan.)	Australis	Dana	Illawarra.	
Anarthrocanna	Australis	Dana		
Cystomeirites (?)	rigida	Dann	Newcastle.	Resembles C. *nutans* (Stern).
Austrella	tenella	Dana	Newcastle.	
Confervites (?)				

APPENDIX III.

1839–1844.

LIST of specimens of Rocks, Fossils, and Minerals collected by the Rev. W. B. CLARKE, in N.S.W., and sent to the late Prof. Sedgwick to be deposited in the Woodwardian Museum of the University of Cambridge, November, 1844.

Districts represented.	No. from each.	Districts represented.	No. from each.
(c) Wianamatta	271	(c) Muswellbrook	47
(c) Hawkesbury	115	(c) Mount Wingen	34
		(a) North of Liverpool Range to Peel River	54
(c) Prospect Hill	22	(a) New England	33
(c) Piakubaba (Pennant Hills)	33	(c) Page	24
(c) Matavai	5	(c) Gill's Cliff	18
(c) Windsor	1	(c) Cedar Brush	23
(c) Maroota	16	(c) Segenhoe	33
(c) Illawarra	588	(c) Upper Hunter	19
(c) Razor Back; Stone Quarry	6		
(c & t) Merrigang and Sutton Forest	24	(c) Paterson	12
		(c) Lewin's Brook; Allyn River	18
(a) Argyle County	145		
(a & t) Murray „	119	(c) Port Stephens	
(a) Twofold Bay, Maneero	16	(c) Stroud	55
(a) Murrumbidgee	71	(c) Smith's Creek, &c.	
(a) Cox's River, Hartley, &c.	34	(c) Williams River	65
		(c) Irrawang and Arowa	38
(c) Mount York	2		
Bathurst sections	79	(c & a) Clarence River and North of	18
(a & c) Mudgee	80		
(c) Awanba	31		
(c) Mulubimba (Newcastle) and up to Loder's Creek	111	(c) Richmond River and Moreton Bay	51
(c) Hunter River (Lower)	34	Miscellaneous	11
(c) Binjaberri	44		
(c) Harpur's Hill	26	Norfolk Island	6
(c) Wollombi	10		
(c) Darlington	16		
(c) Glendon	75	Total	559
(c) Korinda	38	„	2,012
Total	2,012	Grand total	2,571

In the above list (c) refers to carboniferous rocks; (a) to auriferous; (t) to trap.

APPENDIX IV.

1840.

FOSSILS recorded by M. de Verneuil ("*Bulletin de la Soc. Géol. de France,*" tom. xi., p. 177. Séance, 2 Mars, 1840.)

Genus	Species.	Locality.	Remarks.
Orthoceras	sp.	New Holland	SILURIAN species (de Ver.) from Museum of Nat. Hist., Paris.
Spirifer	Small striated		
Cyathophyllum			
Calamopora	Gothlandica		

In the same paper by M. de Verneuil, "*Sur l'importance de la limite qui sépare le calcaire de montagne des formations qui lui sont inférieures*"—he gives the following, as reported by the officers of "La Bonite," as CARBONIFEROUS species determined by himself, viz. :—

Productus	*pustulosus* (Phill.) near *scabriculus* (Sow.)	Collected Mt. Wellington; New Norfolk; Port Dalrymple; Tasmania.	Identical with Yorkshire species.
Spirifer	n. *trigonalis*		
,,	sp. "dichotomous"		
,,	n. *undulatus* (Sow.)		
,,	oblatus = Terebrat. *lævigatus* (Schlotheim.)		Like those of Visé, Belgium. (?) *S. glaber.*
,,	great smooth sp.		
Great Bivalve			
,, Pecten	new sp.		
Calamopora	,,		

N.B.—In the "*Quarterly Journal of the Geol. Soc. Lon.,*" Vol. 1., p. 407, under the head of "Accounts of certain species of *Silurian* fossils from Hobart's Town, N.S.W."!! the above species are accredited to Mt. Wellington—and, the author adds, " the same species are found in Van Diemen's Land, and besides them a great abundance of *Retepora, Cyathophyllum, Calamopora, Clypeaster,* and *Dentalium,* which are rarely met with in the neighbourhood of Mt. Wellington. All these specimens were collected in the hills of Morambiji to the south of the Blue Mountains, and the beds containing them are partly covered, as at Hobart's Town, with recent lignites."!!!

This curious medley is described as "extracted from the '*Voyage de la Bonite: Géol. et Mineralogie, par M.E. Chevalier, p.* 332.'"

I have little doubt that the Silurian species came from the Murrumbidgee, and the Carboniferous from Tasmania.—W.B.C.

APPENDIX V.

1842–3.

LUDWIG LEICHHARDT'S LIST.

FROM "*Notes on the Geology of parts of New South Wales and Queensland, made in 1842-3, translated by G. H. Ulrich, Esq., F.G.S.; and edited by W. B. Clarke,*" in "*Waugh's Almanac,*" Sydney, 1867 and 1868.

	Genus.	Species.	Locality.
Upper Coal-beds.	Large fern, like acrostichum	....	Newcastle.
	Glossopteris	....	,,
	Equisetum (= Phyllotheca)	....	,,
	Lepidodendron (sent to Jardin des Plantes, Paris).		
	Corallinites	Wiltoni	Newcastle.
Upper Marine.	Equisetum	obtuse striatum	Harpur's Hill.
	Spirifer.......................	abundant	,,
	Pectens.......................	....	,,
	Trochus	....	,,
	Pachydomus..................	....	,,
	Turrilites (doubtful)	....	,,
	Many other shells	....	,,
	Ostrea (doubtful).		
	Fenestellæ	⎫	Glendon.
	Spirifer.......................	⎬ .... ⎨	Kelliman's Creek.
	Trochus		
	Hemicardium	⎭	Bell's Creek.
Lower Marine.	Encrinites	⎫	East of Gwydir.
	Terebratula and other shells.	⎬ .... ⎨	Horton River.
			Carrow Brook.
	Trilobite [Doubtless *Brachymetopus.*—W.B.C.]	⎭	Glennie's Stockyard.
	Lycopodium (?)	....	Huskisson's Creek.
	Lepidodendron	....	Manila Creek & Eulowrie.

APPENDIX VI.

1845.

Mr. J. BEETE JUKES, M.A., F.G.S., F.R.S., accompanied the Rev. W. B. Clarke in a visit to the neighbourhood of Wollongong ; and in addition to four species of plants and thirteen Marine fossils from the River Hunter, belonging to the collection in the Woodwardian Museum, at Cambridge, mentioned the following as occurring at Wollongong. [See "*Notes on the Palæontological Formations of New South Wales*," Q.J.G.S., vol. iii, pp. 241-244, 1847.]

Genus.	Species.
Fossil wood in abundance.	
Stenopora	crinita.
Producta	rugata.
Spirifer	subradiatus.
,,	Stokesii.
,,	avicula.
Pachydomus	carinatus.
,,	ovalis (= globosus. Morr.)
Orthonota	sp. nov.
Pleurotomaria	Strzeleckiana.
Bellerophon	contractus (MS.)

APPENDIX VII.

"CARBONIFEROUS FLORA" of the Upper Coal-Beds overlying Palæozoic Marine Beds.

List by Professor Morris, 1845.

Collected by P. E. de Strzelecki.

Genus—Brongniart's.	Species.	Locality.
Sphenopteris	Section of S. *linearis*	Jerusalem, Tasmania.
,,	lobifolia	Newcastle.
,,	alata. var. *exilis*	,, basin.
Glossopteris	Browniana	
Pecopteris — (alethopteris)— (Schimp).	Australis	Jerusalem basin.
(Cycadopteris) (Schimp)	near odontopteroides	,,
Zeugophyllites	elongatus	,,
Phyllotheca	Australis.	

APPENDIX VIII. "A."

CARBONIFEROUS MARINE FOSSILS examined by Professor Morris, 1845.
Collected by P. E. de Strzelecki.

Genus.	Species.	Locality.
	POLYPARIA.	
Stenopora	Tasmaniensis	Mts. Wellington & Dromedary, Tasmania
,,	ovata	,, ,, ,,
,,	informis	Spring Hill, Tasmania
,,	crinita	Illawarra, N.S.W.
Favosites	Gothlandica	Yass Plains, N.S.W.
Amplexus	arundinaceus	Barber's Creek, N.S.W.
Fenestella	ampla	Spring Hill; Mt. Wellington; Eastern Marshes; Tasmania
,,	internata	Mt. Wellington (Tasmania); Patrick's Plains; Raymond Terrace; N.S.W.
,,	fossula	,, ,, ,,
Hemitrypa	sexangula	Mt. Wellington, Tasmania
	MOLLUSCA.	
Allorisma	curvatum	Illawarra, N.S.W.
Pachydomus	antiquatus	Wollongong ,,
,,	cuneatus	,, ,,
,,	lævis	Illawarra ,,
,,	globosus	Illawarra, N.S.W.; Spring Hill (Tas.)
,,	carinatus	Illawarra ,,
Orthonota	costata	Illawarra ,,
,,	compressa	Spring Hill, Tasmania
Eurydesma	cordata	Illawarra [Lochinvar, N. Railway.—W.B.C.] N.S.W.
Pteriuea	macroptera	Spring Hill (Tas.)
Pecten	Illawarrensis	Illawarra, N.S.W.
,,	limœformis	Eastern Marshes (Tas.)
,,	Fittoni	Mt. Wellington (Tas.)
,,	squamuliferus	,,
	BRACHIOPODA.	
Terebratula	cymbæformis	Raymond Terrace, N.S.W.
,,	hastata	Raymond Terrace and Illawarra, N.S.W.
Spirifer	crebristria	Booral, N.S.W.
,,	Darwinii	Glendon ,,
,,	Tasmaniensis	Eastern Marshes (Tas.)
,,	subradiata	Illawarra; Glendon; N.S.W.; Mts. Dromedary and Wellington, Tasmania
,,	avicula	Eaglehawk Neck (Tas.)
,,	vespertilio	,,
,,	Stokesii	Mt. Dromedary (Tas.)
Productus	brachythœrus	Illawarra; Raymond Terrace (N.S.W.); Eastern Marshes; Mt. Wellington (Tas.)
,,	subquadratus	Mts. Dromedary and Wellington (Tas.)

APPENDIX VIII. "A."—continued.

Genus.	Species.	Locality.
	GASTEROPODA.	
Littorina.........	filosa	Booral, N.S.W.
Turritella	tricincta	"
Platyschisma ...	oculus	Harpur's Hill, N.S.W.
" ...	rotundatum	"
Pleurotomaria...	Strzeleckiana ...	Illawarra and Glendon, N.S.W.
" ...	cancellata.........	Illawarra "
" ...	? conica	" "
	HETEROPODA.	
Bellerophon ...	micromphalus ...	Illawarra, N.S.W.
	PTEROPODA.	
Theca	lanccolata.........	Illawarra, N.S.W.
Conularia	lævigata	" and Raymond Terrace, N.S.W.
	CEPHALOPODA.	
Orthoceras	near. undulatum	Yass Plains, N.S.W.
	CRUSTACEA.	
Bairdia	affinis	Booral, N.S.W.
Cythere	sp.	
Trilobites	small impressions	
	PISCES.	
Icthyodorulites..	....	Booral District, N.S.W.

APPENDIX IX.

COAL MEASURE PLANTS—"Carboniferous of Morris": "Oolite of M'Coy"; Professor M'Coy's List, 1847—collected by W. B. Clarke; see "*Annals Natural History*," vol. xx.

Genus.	Author.	Species.	Author.	Locality.
Vertebraria	Royle ..	Australis ..	M'Coy ..	Mulubimba (Newcastle).
Cyclopteris	Brongn..	angustifolia	" ..	Guntawang.
(=Gangamopt. M'C. 1860.)				
Sphenopteris	Brongn...	alata	Brong. ..	Mulubimba.
(S. Hymenophyllites)...	Schimp. .	(Grandinl)..	Goepp.	
	1. 404.			
"	.. Brongn..	lobifolia ..	Morr	id.
"	.. "	hastata ..	M'Coy ..	id.
"	.. "	Germana ..	" ..	id.
"	.. "	plumosa ..	" ..	id.
"	.. "	flexuosa ..	" ..	id.
Glossopteris		Browniana.	Brongn..	Jerry's Plains and id.
"	"	linearis....	M'Coy ..	Wollongong and ? Arowa.
Zeugophyllites........	"	elongatus ..	Morr	Mulubimba.
Phyllotheca	"	Australis ..	Brongn...	id.
"	"	ramosa	M'Coy ..	id.
"	"	Hookeri ..	" ..	id. Arowa ; Clarke's Hill.

NOTE.—Arowa is below Marine beds—Clarke's Hill in Wianamatta.

APPENDIX X.

PLANTS from WIANAMATTA BEDS, collected by W. B. Clarke, and described by Professor M'Coy, 1847. (See "*Annals of Nat. Hist.*," vol. xx.)

Genus.	Author.	Species.	Author.	Locality.
Gleichenites............ = Pecopteris (Morr.) = Cycadopteris (Schimp) = Pecopteris (Carruthers).	......	odontopteroides..	Morr. ..	Clarke's Hill, near Cobbity.
Odontopteris	Brong. ..	microphylla	M'Coy ..	Do. (not figured.)
? Otopteris .... (? Rhacopteris of Feist).	Lindl. ..	ovata	,, ..	(?) Arowa (doubtful.)
Phyllotheca	Brong. ..	Hookeri	,, ..	Clarke's Hill.
Pecopteris............	......	tenuifolia........	,, ..	id.

APPENDIX XI. "B."

CARBONIFEROUS MARINE FOSSILS, determined by Professor M'Coy, 1847. ("*Annals Nat. Hist.*," vol. xx.)

Genus.	Species.	Locality.
	ZOOPHYTA.	
Stenopora	Tasmaniensis	Darlington.
,, ...	crinita	Wollongong ; Black Head ; Darlington.
,, ...	ovata	Darlington.
Fenestella	ampla	Muree ; Bell's Ck. ; Loder's Ck.
,, ...	fossula	Muree.
,, ...	internata.............	Bell's Ck. ; Darlington.
,, ...	undulata.............	Dunvegan or Burragood.
,, ...	? antiqua	} Korinda.
,, ...	? plebeia	
Glauconome	allied to pluma	Burragood.
Cladochonus......	tenuicollis	id.
? Strombodes	Australis	Wagamee.
Turbinolopsis	bina	Burragood.
Amplexus	arundinaceus..........	Curradulla Creek ; Illawarra ; Shoalhaven.
	CRINOIDEA.	
Tribrachyocrinus.	Clarkei...............	Darlington.
Actinocrinus......	....	Wagamee ; Wollamboola.
	CRUSTACEA.	
Bairdia	curta	Burragood.
Cythere	impressa	id.
Brachymetopus ..	Strzeleckii	id.
Phillipsia	? gemmulifera	id.
	MOLLUSCA.	
Atrypa	cymbæformis (M)	Muree ; Black Head.
,, ...	biundata.............	Black Head ; Korinda ; Lewin's Brook.
,, ...	Jukesii	Burragood.
Spirifera	crebristria (M)	id. ; Trevallyn
,, ...	vespertilio	Black Head ; (Eagle-hawk's Neck, Tas.)
,, ...	calcarata	Burragood.
,, ...	avicula..............	Black Head ; Korinda.
,, ...	Darwinii (M)	Loder's Ck. ; Barraba ; Black Head.
,, ...	subradiata	Muree ; Black Head ; Wollongong ; Darlington.
,, ...	? glabra	Maitland ; Irrawang.

APPENDIX XI. " B."—continued.

Genus.	Species.	Locality.
	MOLLUSCA—continued.	
Spirifera	attenuata	Burragood.
,,	Tasmaniensis (M)	Lewin's Brook.
,,	lata	Lewin's Brook.
,,	duodecimcostata	Wollongong ; Muree.
,,	oviformis	Barraba.
Orthis	striatula	Lewin's Brook.
,,	Australis	id.
,,	spinigera	Burragood.
Productus	antiquatus (reticulatus)	Lewin's Brook.
,,	brachythærus	Loder's Ck. ; Korinda (Muree.)
,,	setosus	Lewin's Brook.
,,	scabriculus	Hall's quarry, Hobart Town.
,,	undulatus	Loder's Creek.
Leptæna	sp. (Hardrensis)	Burragood.
Orbicula	affinis	id.
	LAMELLIBRANCHIATA.	
Pecten	squamuliferus (M)	Wollongong.
,,	ptychotis	Burragood.
,,	sub-5-lineatus	Harpur's Hill.
Avicula	tessellata	Burragood.
Pterinea	macroptera (M)	Port Arthur (Tas.)
Eurydesma	cordata (M)	Harpur's Hill.
Inoceramus	Mitchelli	Glendon ; Wollongong.
Pleurorhyncus	Australis	Wollongong.
Allorisma	curvatum (M)	Darlington ; Wollongong ; Glendon.
Orthonota	compressa (M)	Harpur's Hill.
,,	costata (M)	Wollongong.
Modiola	crassissima	Harpur's Hill.
Pachydomus (M)	carinatus (M)	Wollongong ; Port Arthur (Tas.)
,,	globosus	Wollongong.
,,	gigas	id.
,,	sacculus	Black Head ; Wollongong.
,,	ovalis	Wollongong.
,,	pusillus	id.
? Cardinia	exilis	id.
Notomya (M)	securiformis	id.
,,	clarata	id.
? Pullastra	striato-costata	Burragood.
? Venus	gregaria	Wollongong.
	GASTEROPODA.	
Euomphalus	minimus	Burragood.
Pleurotomaria	subcancellata (M)	Loder's Creek.
,,	Strzeleckiana (M)	Wollongong.
,,	Morrisiana	Black Head ; Muree.
Platyschisma	rotundatum (M)	Harpur's Hill.
,,	oculus (M)	id.
	PTEROPODA.	
Theca	lanceolata (M)	Black Head.
Conularia	lævigata (M)	Harpur's Hill
,,	torta	Muree.
,,	tenuistriata	id.
	CEPHALOPODA.	
Bellerophon	micromphalus (M)	Muree.
	interstrialis	Burragood.
Nautilus	? N. sulcatus	id.

N.B.—In the above list, 'M' signifies new genera, and species formed by Professor Morris ; the *italicised* fossils belong to Professor M'Coy. By comparison of lists "A" and "B" with "C" the progress of discovery since 1845 may be ascertained. (*Appendices VII, XI, XVI.*)

APPENDIX XII.

RECORDED as "DEVONIAN FOSSILS" by Samuel Stutchbury, F.G.S., sometime Geological Surveyor in New South Wales, 1851-53.

Genus.	Species.	Locality.	Reference to Reports.
Spirifera		Brucedale	12 April, 1851.
Porites			
Stenopora			
Favosites			
Actinocrinites			
Platycrinites		Errowinbang, or Flyer's Creek.	18 July, 1851.
Rhodocrinites			
Cyathocrinites	with "Pentangular Column."		
Spirifer	Stokesii		
Shell (turbinated)			
Corals			
Cyathophyllum			
Favosites		Nubrigan or Baduldura Creeks.	18 Oct., 1851.
Stromatopora			
Porites			
Crinites			
Molluscs			
Porites	near pyriformis		
(= Heliolites)	interstincta	Near Wellington	26 Jan., 1852.
Caunopora	ramosa		
Lepidodendron			
Crinital	fragments	North side of Horton R., near Mogera Creek.	1 July, 1853.
Leptæna			
Bivalve Shell	unknown		
Orthoceras	sp.		
Piscis	fragment jaw (very doubtful).		
Asaphus (?)	sp.		
Calymene	sp.		
Serpula	sp.		
Bellerophon	globatus		
Orthoceras	sp.		
Euomphalus	sp.		
Turbo	sp.		
Orthonota	sp.		
Mytilus	sp.		
Posidonia (?)	sp.		
Avicula	sp.		
Nucula	sp.		
(Others)	sp.		
Orbicula	sp.	Pallal	1 July, 1853.
Productus	sp.		
Leptæna	sp.		
Orthis	sp.		
Spirifer	disjunctus		
"	others		
Atrypa	sp.		
Terebratula	sp.		
Hypothyris	sp.		
Cyathocrinus	sp.		
Portions of Stems	sp.		
Turbinolopsis	sp.		
Favosites (?)	sp.		
Glauconome	bipinnata		
Fenestella	sp.		
Retepora and others	sp.		
Cirrus and another		Near Taoratooka, Canomodine Creek.	12 April, 1852.
Turbinated Shell			

APPENDIX XIII.

PLANTÆ.

I.—Upper Silurian.

Genus.	Author.	Species.	Author.	Describer.	Locality.	Position in other Countries.
Spirophyton	Kays.	cauda Plusiani	De Kon.	De Kon.	Duntroon.	

II.—Devonian.

Genus.	Author.	Species.	Author.	Describer.	Locality.	Position in other Countries.
Lepidodendron	Stern.	nothum	Ung.	Carruth.	Goonoo Goonoo.	
Cyclostigma	Haugh.	near Kiltorkanense	Feist.	Feist.	id.	

III.—Between Upper Devonian and Lower Carboniferous.

Genus.	Author.	Species.	Author.	Describer.	Locality.	Position in other Countries.
Rhacopteris	Sch.	inequilateralis	Goepp.	Feist.	Port Stephens	Carb. limestone. Silesia.
"		intermedia (between transitionis and Machanæ.)	Stur.	Feist.		Coal-beds. Moravia.
Sphenophyllum	Brong.	sp.				
Rhacopteris	Stern.	near inæquilat		Feist.	Smith's Creek	⎫ Considered to represent Heer's
Cyclostigma	Haugh.	Australe	Feist.	id.	id.	⎬ "Ursa Stufe"(1871), near Isld. S. of
Lepidodendron	Stern.			id.	id.	⎭ Spitzbergen; and Kiltorkan beds.

IV.—Carboniferous underlying or embedded in the same rock with Palæozoic Marine fossils.

Genus.	Author.	Species.	Author.	Describer.	Locality.	Position in other Countries.
Bornia	Stern.	radiata	Broug.	Crepin		
Calamites	Suckow.	varians	Germar.	De Kon.		
Lepidodendron	Stern.	n. disldchum	Stern.	Lesq.	Rouchel Brook.	
"		sp.	M'Coy	id.	Gipps Land.	
"		Australe	Stern	De Kon.	Smith's Creek.	
"		Veltheimianum			Booral.	
Sagenaria	Brong.	sp.	id.	Lesq.	Rouchel Bk.	
? Phyllotheca		rimosa			Harpur's Hill.	
Glossopteris	Brong.	primæva	Feist.	Feist.		
"		Clarkei	id.	id.		
"		Brownium		id.		
Pecopteris	id.	precursor			Spring Hill, Tas. (Strz.)	
Schizopteris	id.	odontopteroides	{ De Kon. ? Crep.			
Zeugophyllites	id.	sp.			Muree	Conularia lævigata.
		sp.				

APPENDIX XIV.

DE KONINCK'S UPPER SILURIAN MARINE SPECIES, N.S.W.

PLANTÆ.

Division—THALLOGENEÆ.

Genus.	Author.	Species.	Author.	Locality.	Associated with
Spirophyton (?)	kayser	cauda phasiani	De Koninck.	Duntroon	Cromus *Murchisoni* (De Koninck).

Division—PROTOZOA. *Class*—RHIZOPODA. *Order*—SPONGIDA.

Stromatopora	Goldfuss	striatella	A. d'Orbigny.	Bell River; Tuena and Lime Kilns.	

Class—ACTINOZOA. *Order*—RUGOSA.

Strombodes	Schweigger	diffluens	Milne-Edwards & Haime.	Berudha River	Heliolites *sp.*
Psychophyllum	M-E. & H.	petellatum	Schotheim	Dangelong.	
Cystiphyllum	Rafinesque & Clifford	Siluriense	Lonsdale	Burrawang.	
Omphyma		Murchisoni (?)	M-E. & H.	Burrawang.	
Cyathophyllum	Goldfuss	articulatum	Wahlenberg	Burrawang.	
Rhyzophyllum (?)		(several species) interpunctatum	De Koninck	Rock Flat Creek.	

Order—TUBULOSA.

Au`opora	Goldfuss	fasciculata	De Koninck.	Bell River	Monticulipora *pulchella* (M-E. & H.) Orthis *canaliculata* (Lindstrom). Plasmopora *petaliformis* (Lonsdale).

APPENDIX XIV—continued.

Order—TABULATA.

Genus.	Author.	Species.	Author.	Locality.	Associated with
Syringopora	Goldfuss	serpens (?)	Linnaeus	Deleget River	Strophomenes pecten (Linn.) Atrypa reticularis (Linn.)
Halysites	Fischer de Waldheim	escharoides	Lamarck	near Wellington.	
Monticulipora	A. d'Orb.	(?) Bowerbanki	M.-E. & H.	Rock Flat Creek.	
"	"	pulchella	M.-E. & H.	Bell River.	
Alveolites	Lam.	repens	Fougt.	Burrawang.	
"	"	raja	De Koninck	Burrawang.	
Striatopora	Hall	Australien	De Koninck	near Yass.	
Favosites	Lam.	cristata	Blumenbach	Burrawang.	
"	"	Forbesi	M.-E. & H.	Borée ; Cabalanine.	
"	"	aspera	A. d' Orb.	Burrawang.	
"	"	multipora (?)	Lonsdale	Limekilns ; Yarralumla.	
"	"	fibrosa	Goldfuss	Burrawang.	
"	"	Gothlandica	Fougt	Yass (Hatton's corner) (Limestone Creek, near Bowning) (Berringullen).	
Propora	M.-E. & H.	tabulata	Lonsdale	Bell River.	
Plasmopora	Dana	petaliformis	"	Bell River.	
Heliolites	"	megastoma	M'Coy	Burrawang.	
"	"	Murchisoni	M.-H. & E.	Burrawang.	

Division—MOLLUSCOIDEA. Class—BRACHIOPODA.

Chonetes	Fisch de Wald	striatella	Dalman	Quedong.	
Leptaena	Dalman	quinquecostata	M'Coy	Yarralumla.	Strophomenes pecten (Linn.) ; Retzia Salteri (Davidson) ; Atrypa hemispherica (Sow.).
"	"	compressa	Sowerby	Duntroon.	
Strophomenes	Rafinesque	pecten	Linn.	Yarralumla ; Duntroon ; Dangelong.	
"	"	rhomboidalis	Wilcken	Rock-flat Creek.	

APPENDIX XIV—continued.

Division—MOLLUSCOIDEA. Class—BRACHIOPODA—continued.

Genus.	Author.	Species.	Author.	Locality.	Associated with
Strophomenes	Rafinesque	fiosa ?	Sowerby	Murrumbidgee; near Yass; Yarralong.	
Pentamerus	Sowerby	Knightii	,,	(Calalamine. (Canobolas.	Stropho-
,,	,,	oblongus	,,	Duntroon	{ Atrypa hemispherica (Sowerby); menes compressa (Sowerby). Monticulipora pulchela (M-E. & H.)
Atrypa	Dalman	orchis canaliculata	Lindstrom	Bel River	
,, (?)	,,	reticularis	Linn.	Duntroon.	
Retzia	King	hemispherica	Sowerby	Duntroon.	
Spirifer	Sowerby	Salteri	Davidson	Yarralumla.	
Meristella	Hall	crispus	Hisinger	Dangelong, Rock Flat Ck.	Many other species.
		tumida (?)	Dalman	Slaughter-house Creek.	

Division—MOLLUSCA. Class—LAMELLIBRANCHIATA.

| Pterinea | Goldfuss | ampliata | J. Phillips | Dangelong | Cronus Bohemicus (Barrande.) Atrypa (?) hemispherica (Sowerby). |
| ,, | | pumila | De Koninck | Yarralumla | Atrypa (?) hemispherica (Sowerby). |

Class—GASTEROPODA. Order—PROSOBRANCHIATA.

Euomphalus	Sowerby	solarioides	De Koninck	Rock Flat Creek.	
,, (?) Subgen. (Omphalotrochus)	Meek	pleurophorus	,,	Murrumbidgee, near Yass.	
		Clarkei	,,	Yarralong, near Yass.	
Bellerophon	Montfort	Jukesii	,,	Rock Flat Creek.	

Class—PTEROPODA. Section—THECOSOMATA.

| Conularia | Miller | Sowerbyi | Defrance | Rock Flat Creek. | |

APPENDIX XIV—continued.

Class—CEPHALOPODA. Order—TETRABRANCHIATA.

Genus	Author	Species	Author	Locality	Associated with
Orthoceras	Breyn	Ibex	Sowerby	Rock Flat Creek	Conularia Sowerbyi.

Class—CRUSTACEA. Order—OSTRACODA.

Genus	Author	Species	Author	Locality	Associated with
Eatonia	Rup. Jones	pelagica	Barrande	Yarralumla	Numerous fragments of trilobites. Alveolites repens (Fougt.).

Class—CRUSTACEA. Order—TRILOBITE.

Genus	Author	Species	Author	Locality	Associated with
Illænus	Dalman	Wahlenbergi	Barrande	Boree Cavern.	Cromus Bohemicus (Barrande and others).
Staurocephalus	Barrande	Clarkei	De Koninck	Rock Flat Creek	Strophomenes pecten (Linn).
Chelrurus	Beyrick	insignis	Beyrick	Yarralumla	Cromus Bohemicus (Barrande).
Encrinurus	Emmrick	punctatus	Brünnich	Yass: Silver Vale. Duntroon	Pentamerus oblongus (Sowerby). Cyathophyllum articulatum (Wahlenberg).
"	Barrande	Barrandei	De Koninck	Yarralumla	Calymene Blumenbachii (Brongn).
Cromus	"	Bohemicus (?)	Barrande	Yarralumla	Similarly to same species at Tobolka, Wohroda, &c.
"	"	Murchisoni	De Koninck	Yarralumla, Quedong	Spirophyton cauda phasiani (De K.).
Calymene	Brongn	Blumenbachii	Brongn	Yarralumla and S. Rwy., 100½ m. near Bowning (1877)	Crustacea referred to above.
Proetus	Steininger	Stokesii (?)	Murchison	Yarralumla	
Lichas	Dalman	nr. "palmata"	Barrande	Rock Flat Creek.	
Bronteus	Goldfuss	Partschi	Barrande	Boree Cavern.	Bronteus Partschi (Barrande).
"	"	geniopeltis	De Koninck	Rock Flat Creek.	
Harpes	"	ungula	Sternberg	Boree Cavern.	

APPENDIX XV.

DE KONINCK'S DEVONIAN SPECIES.

Genus.	Author.	Species.	Author.	Locality.	Associated with

Division—PROTOZOA. *Class*—RHIZOPODA. *Order*—SPONGIDA.

| Archeocyathus? | Billings | Clarkei | De K. | neighbourhood of Yass | Leptæna nobilis (M'C.) |

Division—COELENTERATA. *Class*—ACTINOZOA. *Order*—RUGOSA.

Phillipsastrea	d'Orb.	Verneuili	E. & H.	Cope's Gully.	
Campophyllum	E. & H.	flexuosum	Goldf.	Quedong.	Chaetetes Goldfussi (E. & H.)
Cyathophyllum	Goldf.	vermiculare		Calalamine; Yarralumla.	
"	"	obtortum	E. & H.	Mowara Ck.; in Yass; Quedong.	
"	"	Damnoniense	Lons.	Yarralumla; Cope's Gully.	
"	"	helianthoides	Goldf.	Murrumbidgee; Yass River.	
"	"	several species			
Amplexus	Sow.	Selwyni	De K.	Yarradong; Quedong.	
Coenites	Eich.	expansus	"	Yarradong.	
Billingsia	De K.	alveolaris	"	"	
Syringopora	Goldf.	auloporoides	"	Mowara.	Alveolites subæqualis (M. & H.)
Alveolites	Lam.	obscurus	"	Quedong; Yarradong.	Syringopora aulop. (de K.)
"	"	subæqualis	E. & H.	Mowara	
Favosites	"	Goldfussi	d'Orb.	Yarralumla; Yass Plains; Yarradong; Tuena; Limekilns.	crinoidal fragments.
"	"	basaltica	Goldf.	Shoalhaven gullies; Calalamine.	
"	"	alveolaris	"	near Kempsey	
"	"	polymorpha	"	Quedong, near Yass; Murrumbidgee R., Mowara; Limekilns.	
"	"	reticulata	Blainv.	Quedong	Heliol. porosa (Goldf.)
Heliolites	Dana	fibrosa	Goldf.	Mowara.	Favosites fibrosa (Goldf.)
"	"	porosa	"		

Division—MOLLUSCOIDEA. *Class*—BRACHIOPODA.

Discina	Lam.	Alleghania	Hall.	Yarradong	Murchisonia angulata (d'Arch & V.)
Strophalosia	King	productoides	Mur.	Kempsey; Deringullen Ck.	Chonetes Hardrensis (Phill.)
Chonetes	Fisch. de Wal.	Hardrensis	Phill.	Yass River; Kempsey.	

APPENDIX XV—continued.

Genus.	Author.	Species.	Author.	Locality.	Associated with
Division—MOLUSCOIDEA.				*Class*—BRACHIOPODA.—*Continued*.	
Chonetes	Fisch. de Wal.	coronata	Conrad	Kempsey	
Orthis	Dahm.	interlineata	Sow.	Yarradong	
,,	,,	striatula	Schlo.	Allyn River.	
Leptæna (?)	,,	interstrialis	Phill.		Fenestella (sp.)
,,	,,	nobilis	M'C.	Yarradong, Yass River	Crinoid nr. Rhodocrinus crenatus (Goldf.)
,,	,,	subæquicostata	De K.	Yarradong.	
Pentamerus	Sow.	pumilus			Spirifer, nr. S. Paillettii (De Vern.)
Rhynconella	F. de W.	pleurodon	Phill.	Turon *above* Sofala. Mt. Lambie; Mt. Walker; near Gowrie; Cunningham's Ck.; Clear Ck.; near Bathurst; near Lower Moruya R.	sp. *disjunctus* (Sow.)
Atrypa	Dalm.	pugnus ?	Martin	Yarradong	
,,	,,	reticularis	Linn.	Kempsey; Yarradong.	Orthis *interlineata* (Sow.)
,,	,,	desquamata	Sow.	Yarradong.	
,,	,,	plicatella	De K.		
Spirifer	Sow.	disjunctus	Sow.	Mt. Lambie; Mt. Walker; Turon *ab* Sofala; Colo Colo; Collins's Flat (Bungonia.)	
,,	,,	multiplicatus	De K.	Yarradong.	
,,	,,	cabedanus	De Vern.		
,,	,,	Yassensis	W.B.C. (De K.)	,, and Duntroon ?	
,,	,,	lævisinuatus	De K.	Murrumbidgee; Yass.	
,,	,,	nudus	Sow.	Yarradong.	Atrypa *plicatella* (de K.)
Division—MOLLUSCA.				*Class*—LAMELLIBRANCHIATA.	
Parmeyclas	Hall	ellipticus	Phill.	Yarradong.	
Conocardium	Bronn	Sowerbyi	De K.	,,	
Tellinomya	Hall	Clarkei		,,	
Avicnlopecten	M'Coy	Etheridgei			
,,		Clarkei		Kempsey	another sp. and Atrypa *ret.* (Linn.)
,,		M'Leayi		Macleay River	
Pterinea	Goldf.	laminosa		Yarradong.	

APPENDIX XV—*continued.*

Genus.	Author.	Species.	Author.	Locality.	Associated with
Class—GASTEROPODA.					
Dentalium	Linn.	antiquum	Goldf.	Yarradong; Yass R.	many Spir. *Vassensis* (W.B.C.)
"	"	tenuissimum	De K.	Yarradong.	
Bellerophon	Montf.	convolutus	"	"	
Murchisonia	d'Arch & Vern.	Verneuiliana	"	"	Discina *Alleghania.*
"	"	turris	"	"	
"	"	granifera	"	"	
Pleurotomaria	Defrance	subconica	"	"	
Euomphalus	Sow.	nodulosus	"	"	
"	"	Bigsbyi	"	"	
Loxonema	Phill.	Anglicum	D'Orb.	"	
"	"	antiquum	Münster	"	
"	"	Hennahii	Sow.	"	
Niso (?)	Risso.	deperditum	Goldf.	"	
Mitchellia	De K.	Darwinii	De K.	"	
Natica	Lam.	striatula	"	"	
"	"	cirriformis	Sow.	"	
Class—CEPHALOPODA. *Order*—TETRABRANCHIATA.					
Goniatites	De Haan	Woodsii	De K.	? Yarralumla.	
Cyrtoceras	Goldf.	textile	"	Yarradong.	
Orthoceras	Brug.	subdimidiatum	Münster	Yass River; Murrumbidgee.	
"	"	lineare	"	"	
"	"	sp.		"	
"	"	"		"	

APPENDIX XVI. "C."

DE KONINCK'S CARBONIFEROUS SPECIES.

Genus.	Author.	Species.	Author.	Locality.	Associated with
PLANTÆ.					
Lepidodendron	...	Veltheimianum	Storn.		
Hornia	...	radiata	Brong.		
Calamites	...	varians	Germar.		
Class—POLYPI. Order—ZOANTHARIA. Section—RUGOSA.					
Axophyllum (?)	E. & H.	Thomsoni	De K.	Jervis Bay; Colo Colo............	named after Jas. Thomson, Glasgow.
Lithostrotion	Llwyd.	Irregulare	Phill.	Piper's Ck.; Manning R.	
"		basaltiforme	Conyb. & Phill	Murrumbidgee.	
Cyathophyllum	Goldf.	invernum	De K.	Colo Colo.	
Lophophyllum	E. & H.	minutum	"	Burragood (=Dunvegan), Paterson River.	
"		corniculum		Colo Colo.	
Amplexus	Sow.	arundinaceus (?)	Lons.	Colo Colo; also Barber's Ck. (Strz.) Curadulla (M'C.)	Griffithides Eichwaldi (Fisch). Cladochonus brevicollis (M'Coy.)
Zaphrentis	Rafinesque and Clifford	Phillipsi	E. & H.	Allalong; Colo Colo.	
"	"	Gregoryana	De K.	Jervis Bay; Colo Colo.	
"	"	cainodon	"	Colo Colo & Burragood.	
"	"	robusta	"	Burragood.	
"	"	minuta	"	"	
Cyathaxonia	Michelin				
Order—TUBULOSA.					
Cladochonus	M'Coy	tenuicollis	M'C.	Burragood.	
Section—TABULATA.					
Syringopora	Goldf.	reticulata	Goldf.	Muree; Icthyodorulite Range; Karua. Burragood.	Productus semireticulatus.
"	"	ramulosa (?)	"	Burragood.	
Favosites	Lam.	ovata	Lons.	Glen William; Burragood; Darlington; Singleton Bridge; Harpur's Hill; also Mt. Wellington and Mt. Dromedary, and Norfolk Plains (Tas.) (Strz.)	

APPENDIX XVI. "C."—continued.

Genus.	Author.	Species.	Author.	Locality.	Associated with
Class—ECHINODERMATA. *Order*—CRINOIDEA.					
Symbathocrinus	Phill.	ogivalis	De K.	Burragood.	
Poteriocrinus	Miller	tenuis (?)	Austin	,,	
,,	,,	radiatus (?)	,,	,,	
Actinocrinus	,,	polydactylus	Miller	Glen William.	
Platycrinus	,,	near lævis	,,	Burragood ; and between Hunter and Rouchel Rs. ; Glen William.	
Tribrachyocrinus	M'Coy	Clarkei	M'Coy	Muree ; Darlington.	
Cyathocrinus	Miller	Konincki	W.H.C.	Osterley ; Muree ; Illawarra	See *pentadia corona* (Dana).
Order—ASTEROIDEA.					
Palæaster	Hall	Clarkei	De K.	Russell's Shaft (wet coal).	
Class—BRYOZOA. *Order*—CYCLOSTOMATA.					
Penniretepora	d'Orb	grandis (?)	M'C.	Burragood.	Spirifer *glaber* (Martin.)
Dendricpora	De K.	Hartyi	W.B.C.	id and Pallal	Fenestella *Morrisii* (McC.)
Fenestella	Lons.	plebeia = fossula	M'C.	Glendon ; Muree ; Glen William ; Wingham ; Manning R. ; Singleton ; also Mt. Wellington (Tas.), Gympie (Queens.) [Etheridge.]	Griffithides *Eichwaldi* (Fisch).
,,	,,	propinqua	De K.	G'en William, Pallal, Glendon Brook.	Rhynoconella *pleurodon* (Phill.)
,,	,,	multiporata	M'C.	Burragood	
,,	,,	internata	Lons.	Patrick's Plains ; Muree ; Glendon ; Icthyodorulite Range. [Mt. Wellington (Tas.)]	
,,	,,	Morrisii	M'C.	Burragood.	Rhyn. *pleurodon* (Phillips), Crinoidal fragments.
Protoretepora	De K.	gracilis (?)	Dana.	Burragood, Glendon.	
,,	,,	ampla	Lons.	Muree, and between that and Morpeth ; Bell's Ck. ; Loder's Ck. ; Glendon Stony Ck. ; xx-mile cutting N. Rw. also Spring Hill, Mt. Wellington, and Eastern Marshes, Tas. (Strz.)]	Productus *brachythærus* (Sow.)
Retepora (?)	Lam.	laxa	De K.	Colo Colo ; Burragood	Orthis *resupinata* (Mart.)
Polypora	M'C.	papillata (with 2 specks of coal)	M'C.	near Buchan, Gloucester R.	Orthotetes *crenistria* (Phill.)

APPENDIX XVI. "C."—continued.

Genus.	Author.	Species.	Author.	Locality.	Associated with
Class—BRACHIOPODA.				*Family*—PRODUCTIDÆ.	
Productus	Sow.	Cora	d'Orb.	Tillegharry and Karua R. to Dungog. S.E. of Buchan Mountain.	
,,	,,	magnus	Meek and Worthen.		many other species.
,,	,,	semireticulatus	Martin	Glen William and Colo Colo.	
,,	,,	undatus	Defrance	Paterson R.	
,,	,,	Flemingii	Sow	Buchan; Glen William; Burragood; Colo Colo; Harpur's Hill; Lewin's Brook; Don River, Queensland.	
,,	,,	punctatus	,,	Buchan; Gloucester R.; Karua and Williams Rivers	
,,	,,	fimbriatus	,,	Tillegharry; Karua and Williams Rivers	spirifer duplicostatus (Mill.)
,,	,,	scabriculus	Mart.	Williams River	
,,	,,	brachythaerus	Sow.	Loder's Ck.; Korinda; Wollongong Point; Raymond Terrace; Ellalong; Darlington; Muree to Morpeth.	
,,	,,	fragilis	Dana	Wollongong Point; Braxton.	
,,	,,	Clarkei	Ether. (senr.)	Branxton; Burragood (Bowen River, Queensland).	
,,	,,	aculeatus	Mart.	Colo Colo; Pallal; Gwydir River.	
Chonetes	Fisch.	papilionacea	Phill.	Dungog; Underbank, in oolite like limestone grit.	
,,	,,	Lagucsslani	De K.	Burragood; Pallal; Colo Colo; Karua to Dungog. [Cracow, Queensland.]	
Family—STROPHOMENIDÆ.					
Strophomena	Raf.	analoga	Phill.	Branxton; between Page and Bouchel Rivers; Burragood; Colo Colo; between Karua and Dungog. [Gympie, (Ether. senr.)]	
Orthotetes	Fisch.	crenistria	Phill.	Colo Colo; between Karua and Dungog.	
Orthis	Dalman.	resupinata	Mart.	Lewin's Brook; Pallal; Burragood; Colo Colo.	Prod. semireticulatus (Martin).
,,	,,	Michelini	Leveillé	Buchan; Burragood; Colo Colo.	

APPENDIX XVI. "C."—continued.

Genus.	Author.	Species.	Author.	Locality.	Associated with
Family—RHYNCONELLIDÆ.					
Rhynconella	Fisch.	pleurodon	Phill.	Coyeo (Page River—over coal); between Karua and Dungog; Burragood; Tillegharry.	
,,	,,	inversa	De K.	Muree.	
Family—SPIRIFERIDÆ.					
Athyris	M'Coy	planosulcata	Phill.	Burragood; Underbank.	
Spirifer	Sow.	lineatus	Mart.	Buchan; betw. Karua and Dungog; Muree; Colo Colo.	[or *S. lineatus* (Mart.), varietas?]
,,	,,	crebristria	Morr.	Booral; Trevallyn; Burragood	
,,	,,	glaber	Mart.	Ællalong; Stony Ck.; Harpur's Hill; Muree; Brauxton; Morpeth; Mt. Wingen (over coal); Muniwarree (Black Head); Anvil Ck.; Greta; Darlington Burragood; Glen William; N. Rwy. near Lochinvar and to Maitland Glendon. [Thas.: Mts. Wellington and Dromedary.]	
,,	,,	Darwinii	Mor.	Glendon; Loder's Creek; Barraba; Black Head; Maitland; Muree; xx-mile cutting, N. Rwy.; Harpur's Hill; Wingen.	
,,	,,	oviformis	M'C.	Burragood; Trevallyn; Ællalong.	
,,	,,	duodecimcostatus.	M'C.	Wollongong; Stroud; Bomhaderra Ck.; Muree; Ællalong.	
,,	,,	Strzeleckii	De K.	Maitland to 34 miles from Newcastle N. Rwy.; Muree. (Gympie, *Ether. Senr.*)	
,,	,,	Clarkei	De K.	Wollongong; Jervis Bay.	
,,	,,	pinguis	Sow.	Burragood; Glen William; Karua.	

APPENDIX XVI. "C."—*continued.*

Genus.	Author.	Species.	Author.	Locality.	Associated with
Family—SPIRIFERIDÆ—*continued.*					
Spirifer	Sow.	convolutus =avicula	Phill. Sow. and Morr.	Muree; Russell's Shaft (over coal); Stony Ck.; Anvil Ck.; Branxton; St. Helier's; Coyeo; Mt. Wingen; Allalong; Wollongong; Gimbela; N. Hwy.; Black Head. [Eagle Hawk Neck, Tas.]	S. *vespertilio* (Sow.)
,,	,,	vespertilio	Sow.	Excepting Wollongong and Gimbela, same as S *convolutus.* Lewin's Brook; Colo Colo.	S. *convolutus* (Phill.)
,,	,,	latus	M'C.	Stroud.	
,,	,,	triangularis	Mart.	Muree; Branxton; St. Helier's; Burngool; Colo Colo; Mulberring Ck. Allalong; Tillogharry; Glen William; Cedar Brush (Serenhoe); Underbank Jervis Bay. [Bowen R., Queensl. Either, Senr.]	Terebratula *sacculus* (Mart.)
,,	,,	bisulcatus	Sow.		
,,	,,	Tasmaniensis	Morr.	Muree; Gret (over coal); Korinda Allalong; Mt. Wingen (over coal) Coyeo (ditto); Nowra Hill (Shoalhaven R.) [Eastern Marshes and Mt. Wellington, Tas., *Morris.*] Colo Colo; Glen William.	
Spiriferina	D'Orb.	exsuperans	De K.	,,	
,,	,,	cristata	Schlot.		
,,	Davidson	insculpta	Phill.	? Murrumbidgee.	
Cyrtina	Lhwyd	septosa (? drifted)	Phill.	Harpur's Hill, Stroud.	
Terebratula	,,	sacculus (var.)	Mart.	Muree; Stroud; Icthyodorulite Range; between Hunter and Rouchel Rivers Black Head; Harpur's Hill; Korinda Lewin's Brook; Lochinvar; Pallal Burragood.	sp. *trigonalis* (Mart.)
,,	,,	hastata (var.)	Sow.		

APPENDIX XVI. "C."—continued.

Division—MOLLUSCA. Class—LAMELLIBRANCHIATA.

Genus.	Author.	Species.	Author.	Locality.	Associated with
Scaldia (?)	De Ryck.	depressa	De K.	Buchan.	
"	"	lamellifera	"	Harpur's Hill.	
Sanguinolites	M'C.	undatus	Dana.	Wollongong; Mt. Vincent; Burragood; Muree to Morpeth.	
"	"	Mitchelli	De K.	Æhaloug; Wollongong.	
"	"	Etheridgii	"	Mt. Vincent; Muree to Morpeth.	
"	"	M'Coyi	"	Wollongong.	
"	"	curvatus	Morr.	Mt. Gimbela; Wollongong; Darlington Glendon.	
"	"	Tenisoni (after Tenison-Woods)	De K.	Burragood.	*T. semireticulatae* (Mart.) crinoidal fragmts.
Clarkia	De K.	myiformis	Dana.	Gimbela, Wollongong.	
Cardiomorpha	"	gryphoides	De K.	Stony Ck.	
Edmondia (?)	"	striatella	M'C.	Ichthyodorulite Range.	Pleurotomaria flosa.
"	"	striato-costata	De K.	Burragood.	
"	"	nobilissima	M'C.	Muree to Morpeth	
"	"	intermedia	M'C.	Wollongong; Ichthyodorulite Range	
Cardinia	Agass.	exilis	Sow.		Pleurotomaria *Morrisiana* (M'C.) and other fossils.
Pachydomus	Morr.	globosus	"	Wambo; Wollongong; Lochinvar; Spring Hill (Tas.)	P. *globosus.*
"	"	lævis	M'C.	Illawarra; Harpur's Hill; Lochinvar.	
"	"	gigas	"	Wollongong; Mt. Vincent.	
"	"	ovalis	Dana	Wollongong.	
"	"	cyprinus	M'C.	"	
"	"	pusillus	Dana.	" and Black Head.	
"	"	politus	De K.	" and Jervis Bay.	
Meonia	Dana.	Danai	W.B.C.		
"	"	Konincki	Dana.	Calamine.	[Astartila (Dana)].
"	"	elongata	"	Black Head; Wollongong.	
Pleurophorus	"	gracilis	De K.	Wollongong; Muree.	
"	King	Morrisii	Morr.	"	
"	"	biplex			
"	"	carinatus (?)		"	[Orthonota (?) *costata* (Morr.)].

APPENDIX XVI, "C"—continued.

Division—MOLLUSCA. Class—LAMELLIBRANCHIATA—continued.

Genus.	Author.	Species.	Author.	Locality.	Associated with
Conocardium	Bronn	Australe ?	M'C.	N. Rwy., Maitland to Stony Creek.	
Tellinomya	Hall	Darwinii	De K.	Burragood.	
Palæarca	,,	costellata	M'C.	,,	*V. costellata*
,,	,,	interrupta	De K.	,,	
Mytilus	,,	sub-argata	Phill.		
,,	Linn	crassiventer	De K.	Branxton.	
Aviculopecten	M'C.	Bigsbyi	Dana.	Gimbela ; Cooloongatta.	
,,	,,	lentusculus	M'C.	Harpur's Hill ; Wollongong ; Muree.	
,,	,,	sub-quinque-lineatus	Morr.	East. Marshes ; Muree ; Wollongong.	
,,	,,	limæformis	,,	Burragood.	
,,	,,	consimilis	,,		
,,	,,	dequilis	,,	Wollongong.	
,,	,,	elongatus	,,	Burragood and Wollongong.	
,,	,,	ptychotis	,,	Icthyodorulite Range.	
,,	,,	Knockonniensis	De K.	Burragood.	
,,	,,	Hardyi	M'C.	Karua to Dungog.	
,,	,,	cingendus	Sow.	Duguld's Hill ; Burragood	Brachymetopus
,,	,,	granosus	M'C.	Burragood.	(M'Coy)
,,	,,	Forbesi	Phill.		
,,	,,	tessellatus	De K.	Iase Harpur's Hill, N. Rwy.	
,,	,,	profundus	Morr.	Muswellbrook ; Muree to Morpeth ; [Mt. Wellington, Tas.]	
,,	,,	Fittoni		Illawarra ; Harpur's Hill.	
Aphanala	De K.	Illawarrensis	M'C.	Glendon ; Wollongong ; Muree ; Branxton ; Anvil Ck.	
,,	,,	Mitchelli		Branxton.	
Pectinea	Go.df.	giganteu	De K.	Maitland ; Spring Hill (Tas.)	
Avicula	,,	unacroptera	Morr.	Glen William.	
,,	Linn	lata	M'C.	Muree.	
,,	,,	sublunulata	De K.	Burragood.	*Strzelecki*[1]
,,	,,	Hardyi	,,	Harpur's Hill.	
,,	,,	decipiens	,,	,,	
,,	,,	lntumescens	,,		

142 *Sedimentary Formations*

APPENDIX XVI. "C"—*continued*.

Genus.	Author.	Species.	Author.	Locality.	Associated with
Class—PTEROPODA.					
Conularia	Miller	tenuistriata	M°C.	Muree.	
,,	,,	quadrisulcata	Miller	Buchan.	
,,	,,	lævigata	Morr.	Harpur's Hill; Black Ck.; Muree	*C. tenuistriata* (M°C.) plants among which is Schizopteris.
,,	,,	inornata	Dana	Glendon; xx-mile cutting, N. Rwy.	
Class—GASTEROPODA. *Order*—PROSOBRANCHIATA.					
Dentalium	Linn.	cornu	De K.	Karua; Icthyodorulite Range.	
Platyceras	Conrad	angustum	Phill.	Coyeo; Burragood; Rouchel.	
,,	,,	trilobatum	Dana	,,	
,,	,,	altum	,,	Harpur's Hill; Darlington; Pallal.	
,,	,,	tenella	Mart.	Harpur's Hill; Colo Colo.	
Porcellia	Leveille	Woodwardii	M°C.	Burragood.	
Pleurotomaria	Defrance	Morrisiana	Morr.	Munniwarree, Muree.	
,,	,,	subaucellata	Sow.	Muree; Loder's Ck.; Illawarra	
,,	,,	striata	Phill.	Burragood; Duguid's Hill.	
,,	,,	gemmulifera	De K.	xx-mile, N. Rwy.	
,,	,,	humilis	,,	Muree.	
,,	,,	naticoides	,,	Harpur's Hill.	
,,	,,	helicineformis	,,	Burragood.	
Murchisonia	d'Arch. & Vern.	trifilata	Dana	Harpur's Hill; Muree; Wollongong.	*M. trifilata* (in great quantity), Dana.
,,	,,	Vernculiana	De K.	Munniwarree.	
Euomphalus	Sow.	oculus	Sow.	Harpur's Hill.	
,,	,,	minimus	M°C.	Burragood.	Palæarea *sp.*
,,	,,	catillus	Martin	Tillegharry.	
Macrocheilus	Phill.	filosus	Sow.	Booral, Icthyodorulite Range	
,,	,,	acutus	De K.	Burragood.	
Loxonema	,,	difficilis	,,	Karua River; and Icthyodorulite Range.	
,,	,,	constricta	Martin	,,	
,,	,,	acutissima	De K.	,,	
,,	,,	rugifera	Phill.	Burragood.	

APPENDIX XVI. "C"—continued.

Genus.	Author.	Species.	Author.	Locality.	Associated with

Class—CEPHALOPODA. Order—TETRABRANCHIATA.

Genus	Author	Species	Author	Locality	Associated with
Goniatites	De Haan	micromphalus	Morr.	Illawarra; Muree.	
"	"	strictus	Dana	Wollongong; Harper's Hill.	
Orthoceras	Breynius	striatum	Sow.	Wollongong; Stony Creek (Russell's Shaft).	
"	"	Martinianum (?)	De K.	Ichthyodorulite Range.	
Cameroceras	Conrad.	Phillipsii	"	Glen William.	S. bisulcatus (Sow.); Chonetes papilionacea; Orthis resupinata (Martin)
Nautilus	Breynius	subsulcatus	Phill	Karua.	

Class—CRUSTACEA. Order—OSTRACODA. Section—ENTOMOSTRACA.

Genus	Author	Species	Author	Locality	Associated with
Polycope	Sow.	simplex	Jones & Kirkby	Muree.	
Entomis	Jones	Jonesii	De K.	Pumly; Muree.	Entomis

Order—TRILOBITA.

Genus	Author	Species	Author	Locality	Associated with
Phillipsia	Portlock	seminifera	Phill.	Colo Colo; Burragood.	
Griffithides	"	Eichwaldi	Fisch.	Upper Williams River.	
Brachymetopus	M'C.	Strzeleckii	M'C.	Burragood, Glen William.	

Division—VERTEBRATA. Class—PISCES. Order—CLASMOBRANCHII.

Genus	Author	Species	Author	Locality	Associated with
Tomodus	Agass.	convexus?	Agass.	Tillegharry.	

APPENDIX XVI. "C."—continued.

RÉSUMÉ GÉOLOGIQUE.

Le travail qui précède comprend la description de cent soixante-seize espèces de fossiles carbonifères qui toutes ont été recueillies par les soins du révérend W. B. Clarke dans toute l'étendue de la Nouvelle-Galles du Sud et dont la plupart ont été figurées avec la plus grande exactitude possible.

Parmi ces espèces, on en compte cent et trois dont l'existence n'a pas encore été signalée en Australie, cinquante-neuf qui sont nouvelles pour la science et soixante-quatorze dont la présence a été constatée dans le terrain carbonifère de l'Europe.

Le tableau suivant dans lequel j'ai marqué par un astérisque l'existence de chacune des espèces soit en Europe, soit dans l'une des trois importantes régions de l'Australie, à savoir : la Nouvelle-Galles du Sud, la Tasmanie et la terre de la Reine ou Queensland ([1]), permettra de saisir par un simple coup d'œil leur distribution dans ces diverses contrées.

([1]) Il est assez remarquable que la colonie de Victoria n'ait encore fourni aucun fossile du calcaire carbonifère, quoique les terrains paléozoïques n'y fassent pas défaut.

No. d'ordre.		N.-G. du Sud.	Tasmanie.	Queensland.	Europe.
1	*Axophyllum Thomsoni*, L.-G. de Koninck	*	..	..	..
2	*Lithostrotion irregulare*, J. Phillips	*	..	..	*
3	,, *basaltiforme*, Conybeare et Phillips	*	..	..	*
4	*Cyathophyllum inversum*, L.-G. de Koninck	*	..	..	..
5	*Lophophyllum minutum*, L.-G. de Koninck	*	..	..	..
6	,, *corniculum*, L.-G. de Koninck	*	..	..	..
7	*Amplexus arundinaceus!* W. Lonsdale	*	..	..	..
8	*Zaphrentis Phillipsi*, Milne Edwards et J. Haime	*	..	..	*
9	,, *Gregoryana*, L.-G. de Koninck	*	..	..	..
10	,, *cainodon*, L.-G. de Koninck	*	..	..	..
11	,, *robusta*, L.-G. de Koninck	*	..	..	..
12	*Cyathaxonia minuta*, L.-G. de Koninck	*	..	..	..
13	*Cladochonus tenuicollis*, F. McCoy	*	..	..	..
14	*Syringopora reticulata*, A. Goldfuss	*	..	..	*
15	,, *ramulosa!* A. Goldfuss	*	..	..	*
16	*Favosites crata*, W. Lonsdale	*	*	..	..
17	*Symbathocrinus spiralis*, L.-G. de Koninck	*	..	..	..
18	*Poteriocrinus tenuis!* T. Austin	*	..	..	*
19	,, *radiatus!* T. Austin	*	...	..	*
20	*Platycrinus lævis*, Miller	*	..	..	*
21	*Actinocrinus polydactylus*, Miller	*	..	..	*
22	*Tribrachiocrinus Clarkei*, F. McCoy	*	..	..	..
23	*Cyathocrinus Konincki*, W. B. Clarke	*	..	..	..
24	*Palæaster Clarkei*, L.-G. de Koninck	*	..	..	..
25	*Penniretepora grandis!* F. McCoy	*	..	..	*
26	*Dendricopora Hardyi*, W. B. Clarke	*	..	..	..
27	*Fenestella plebeia*, F. McCoy	*	*	*	*
	Fenestella fossula, W. Lonsdale	*	*	*	..
28	,, *propinqua*, L.-G. de Koninck	*	..	..	..
29	,, *multiporata*, F. McCoy	*	..	..	*
30	,, *Morrisii*, F. McCoy	*	..	..	*
31	,, *gracilis*, J. D. Dana	*	..	..	..
32	,, *internata*, W. Lonsdale	*	..	..	..
33	*Protoretepora ampla*, W. Lonsdale	*	..	..	..
34	*Retepora! laxa*, L.-G. de Koninck	*	..	..	..
35	*Polypora papillata!* F. McCoy	*	..	..	*
36	*Productus Cora*, A. d'Orbigny	*	..	*	*
37	,, *magnus*, F. B. Meek et A. H. Worthen	*	..	..	..

New South Wales. 145

APPENDIX XVI, "C."—*continued.*

No. d'ordre.		N.-G. du Sud.	Tasmanie.	Queensland.	Europe.
38	*Productus semireticulatus*, W. Martin	*	..	..	*
39	,, *Flemingii*, J. Sowerby	*	..	*	*
40	,, *undatus*, Defrance	*	..	..	*
41	,, *punctatus*, W. Martin	*	..	..	*
42	,, *fimbriatus*, J. Sowerby	*	..	..	*
43	,, *scabriculus*, W. Martin	*	..	..	*
44	,, *brachythærus*, G. Sowerby	*	*	..	..
45	,, *fragilis*, J. D. Dana	*	..	..	..
46	,, *Clarkei*, R. Etheridge	*	..	*	..
47	,, *aculeatus*, W. Martin	*	..	..	*
48	*Chonetes papilionacea*, J. Phillips	*	*	*	*
49	,, *Laguessiana*, L.-G. de Koninck	*	..	*	*
50	*Strophomenes analoga*, J. Phillips	*	..	..	*
51	*Orthotetes crenistria*, J. Phillips	*	..	..	*
52	*Orthis resupinata*, W. Martin	*	..	..	*
53	,, *Michelini*, C. Leveillé	*	..	..	*
54	*Rhynchonella pleurodon*, J. Phillips	*	..	..	*
55	,, *inversa*, L.-G. de Koninck	*	..	..	..
56	*Athyris plancsulcata*, J. Phillips	*	..	..	*
	Athyris ambigua ? J. Sowerby	*	..	..	..
57	*Spirifer lineatus*, W. Martin	*	..	..	*
	Spirifer lineatus, var. *crebristria*, J. Morris	*	..	..	*
58	,, *glaber*, W. Martin	*	..	..	*
59	,, *Darwinii*, J. Morris	*	..	..	..
60	,, *subradiatus*, G. Sowerby	*	*	*	..
61	,, *oviformis*, F. McCoy	*	..	..	..
62	,, *duodecimcostatus*, F. McCoy	*	..	*	..
63	,, *Strzeleckii*, L.-G. de Koninck	*	..	..	..
64	,, *Clarkei*, L.-G. de Koninck	*	..	..	*
65	,, *pinguis*, J. Sowerby	*	..	..	*
66	,, *convolutus*, J. Phillips	*	*	*	*
67	,, *vespertilio*, G. Sowerby	*	*	..	..
68	,, *latus*, F. McCoy	*	..	..	..
69	,, *triangularis*, W. Martin	*	..	..	*
70	,, *bisulcatus*, J. Sowerby	*	..	*	..
71	,, *Tasmaniensis*, J. Morris	*	*	..	..
72	,, *exsuperans*, L.-G. de Koninck	*	..	..	..
73	*Spiriferina cristata*, v. Schlotheim	*	..	..	*
74	,, *insculpta*, J. Phillips	*	..	..	*
75	*Cyrtina septosa*, J. Phillips	*	..	..	*
76	*Terebratula sacculus*, W. Martin	*	..	..	*
	Terebratula, var. *cymbæformis*, J. Morris	*	..	..	..
77	*Scaldia ? depressa*, L.-G de Koninck	*	..	..	..
78	,, *lamellifera*, L.-G. de Koninck	*	..	..	..
79	*Sanguinolites undatus*, J. D. Dana	*	..	..	..
80	,, *Mitchellii*, L.-G. de Koninck	*	..	..	..
81	,, *Etheridgei*, L.-G. de Koninck	*	..	..	..
82	,, *McCoyi*, L.-G. de Koninck	*	..	..	..
83	,, *curtatus*, J. Morris	*	..	..	..
84	,, *Tenisoni*, L.-G. de Koninck	*	..	..	..
85	*Clarkia myiformis*, L.-G. de Koninck	*	..	..	..
86	*Cardiomorpha gryphoides*, L.-G. de Koninck	*	..	..	..
87	,, *striatella*, L.-G. de Koninck	*	..	..	*
88	*Edmondia ? striato-costata*, F. McCoy	*	..	..	..
89	,, *nobilissima*, L.-G. de Koninck	*	..	..	..
90	,, *intermedia*, L.-G. de Koninck	*	..	..	..
91	*Cardinia exilis*, F. McCoy	*	..	..	..
92	*Pachydomus globosus*, J. D. Sowerby	*	..	..	..
93	,, *lævis*, J. D. Sowerby	*	..	..	..
94	,, *gigas*, F. McCoy	*	..	..	..
95	,, *ovalis*, F. McCoy	*	..	..	..
96	,, *cyprinus*, J. D. Dana	*	..	..	..

K

APPENDIX XVI, "C."—continued.

No. d'ordre.		N.-G. du Sud.	Tasmanie.	Queensland.	Europe.
97	Pachydomus pusillus, F. McCoy	*	..	..	..
98	,, politus, J. D. Dana	*	..	..	..
99	,, Danai, L.-G. de Koninck	*	..	..	..
100	Mæonia Konincki, W. B. Clarke	*	..	..	..
101	,, elongata, J. D. Dana	*	..	..	..
102	,, gracilis, J. D. Dana	*	..	..	..
103	Pleurophorus Morrisii, L.-G. de Koninck	*	..	..	..
	Orthonota ? costata, J. Morris	*	..	..	..
104	Pleurophorus biplex, L.-G. de Koninck	*	..	..	..
105	,, carinatus ? J. Morris	*	..	..	..
106	Conocardium Australe ? F. McCoy	*	..	..	..
107	Tellinomya Darwini, L.-G. de Koninck	*	..	..	..
108	Palæarca costellata, F. McCoy	*	..	..	*
109	,, interrupta L.-G. de Koninck	*	..	..	..
110	,, subarguta, L.-G. de Koninck	*	..	..	..
111	Mytilus crassiventer, L.-G. de Koninck	*	..	..	..
112	,, Bigsbyi, L.-G. de Koninck	*	..	..	..
113	Aviculopecten leniusculus, J. D. Dana	*	..	..	..
114	,, subquinquelineatus, F. McCoy	*	..	..	..
115	,, limæformis, J. Morris	*	*	..	..
116	,, consimilis, F. McCoy	*	..	..	*
117	,, depilis, F. McCoy	*	..	..	..
118	,, elongatus, F. McCoy	*	..	..	*
119	,, ptychotis, F. McCoy	*	..	..	..
120	,, Knockonniensis, F. McCoy	*	..	..	..
121	,, Hardyi, L.-G. de Koninck	*	..	..	..
122	,, eingendus, F. McCoy	*	..	..	*
123	,, granosus, J. Sowerby	*	..	..	*
124	,, Forbesi, F. McCoy	*	..	..	*
125	,, tessellatus, J. Phillips	*	..	..	..
126	,, profundus, L.-G. de Koninck	*	..	..	..
127	,, Fittoni, J. Morris	*	*	..	..
128	,, Illawarrensis, F. McCoy	*	..	..	..
129	Aphanaia Mitchellii, F. McCoy	*	..	..	..
130	,, gigantea, L.-G. de Koninck	*	..	..	..
131	Pterinea macroptera, J. Morris	*	..	..	*
132	,, lata, F. McCoy	*	..	..	..
133	Avicula sublunulata, L.-G. de Koninck	*	..	..	..
134	,, Hardyi, L.-G de Koninck	*	..	..	..
135	,, decipiens, L.-G. de Koninck	*	..	..	..
136	,, intumescens, L.-G. de Koninck	*	..	..	..
137	Conularia tenuistriata, F. McCoy	*	..	..	..
138	,, quadrisulcata, Miller	*	..	..	*
139	,, lævigata, J. Morris	*	..	..	..
140	,, inornata, J. D. Dana	*	..	..	..
141	Dentalium cornu, L.-G. de Koninck	*	..	..	*
142	Platyceras angustum, J. Phillips	*	..	..	*
143	,, trilobatum, J. Phillips	*	..	..	..
144	,, altum, J. D. Dana	*	..	..	..
145	,, tenella, J. D. Dana	*	..	..	*
146	Porcellia Woodwardii, W. Martin	*	..	..	..
147	Pleurotomaria Morrisiana, F. McCoy	*	..	..	..
148	,, subcancellata, J. Morris	*	..	..	..
149	,, striata, J. Sowerby	*	..	..	*
150	,, gemmulifera, J. Phillips	*	..	..	*
151	,, humilis, L.-G. de Koninck	*	..	..	..
152	,, naticoides, L.-G. de Koninck	*	..	..	*
153	,, helicinæformis, L.-G. de Koninck	*	..	..	..
154	Murchisonia trifilata, J. D. Dana	*	..	..	..
155	,, Verneuiliana, L.-G. de Koninck	*	..	*	*
156	Euomphalus ocutus, J. D. Sowerby	*	..	..	..
157	,, minimus, F. McCoy	*	..	..	..

APPENDIX XVI, "C."—continued.

No. d'ordre.		N.-G. du Sud.	Tasmanie.	Queensland.	Europe.
158	Euomphalus catillus, W. Martin	*	..	..	*
159	Macrocheilus fusus, J. D. Sowerby	*	..	..	..
160	„ acutus, J. Sowerby	*	..	..	*
161	Loxonema difficilis, L.-G. de Koninck	*	..	..	..
162	„ constricta, W. Martin	*	..	..	..
163	„ acutissima, L.-G. de Koninck	*	..	..	..
164	„ rugifera, J. Phillips	*	..	..	*
165	Goniatites micromphalus, J. Morris	*	..	..	..
166	„ strictus, J. D. Dana	*	..	..	..
167	Orthoceras striatum, J. Sowerby	*	..	..	*
168	Martinianum! L.-G. de Koninck	*	..	..	*
169	Camerocerus Phillipsii, L.-G. de Koninck	*	..	..	..
170	Nautilus subsulcatus, J. Phillips	*	..	..	*
171	Polycope simplex, T. R. Jones et J. W. Kirkby	*	..	..	*
172	Entomis Jonesii, L.-G. de Koninck	*	..	..	..
173	Phillipsia seminifera, J. Phillips	*	..	..	*
174	Griffithides Eichwaldi, G. Fischer de Waldheim	*	..	*	*
175	Brachymetopus Strzeleckii, F. McCoy	*	..	..	*
176	Tomodus convexus? L. Agassiz	*	..	..	..
	TOTAUX	176	9	12	74

En ajoutant à la liste qui précède les espèces suivantes qui ne se sont pas trouvées parmi les nombreux échantillons qui m'ont été communiqués par le révérend W. B. Clarke, mais qui ont été décrites par les auteurs dont j'ai cité les ouvrages au commencement de mon travail, on arrivera à un nombre total de deux cent quarante-neuf espèces ([1]).

No. d'ordre.		N.-G du Sud.	Tasmanie.	Queensland.	Europe.
1	Favosites (Stenopora) crinitus, W. Lonsdale	*	..	..	..
2	„ „ Tasmaniensis, W. Lonsdale	*	*	..	..
3	„ „ informis, W. Lonsdale	..	*	..	..
4	Stromboden? Australis, F. McCoy	*	..	..	..
5	Turbinolopsis? bina? W. Lonsdale	*	..	..	..
6	Ceriopora? laxa, R. Etheridge	*	..	*	..
7	Fenestella undulata, J. Phillips	*	..	..	*
8	„ media, J. D. Dana	*	..	..	..
9	Glauconome pluma? J. Phillips	*	..	..	*
10	Hemitrypa sexangula, W. Lonsdale	*	..	..	..
11	Lingula ovata, J. D. Dana	*	..	..	..
12	Discina affinis, F. McCoy	*	..	..	..
13	Siphonotreta? curta, J. D. Dana	*	..	..	*
14	Productus rugatus? J. Phillips	..	..	*	*
15	„ subquadratus, J. Morris	*	..	*	..
16	Spirifer Stokesii, Koenig	..	*	..	..
17	„ paucicostatus, G. B. Sowerby	*	..	..	..
18	Sanguinolites Glendonensis, J. D. Dana	*	..	..	..
19	„ audax, J. D. Dana	*	..	*	..
20	Edmondia? concentrica, R. Etheridge	..	..	*	..
21	„ obovata, R. Etheridge	*	..	..	..
22	Solecurtus? ellipticus, J. D. Dana	*	..	..	..
23	Solecurtus? planulatus, J. D. Dana	*	..	..	..
24	Astarte? gemma, J. D. Dana	*	..	..	..
25	Pachydomus (Astartila) cytherea, J. D. Dana	*	..	..	..

([1]) Je crois devoir faire observer que je ne garantis pas l'exactitude de ces espèces, dont plusieurs me paraissent être fort douteuses.

APPENDIX XVI, "C."—continued.

No. d'ordre.		N.-G. du Sud.	Tasmanie.	Queensland.	Europe.
26	Pachydomus (Astartila) politus, J. D. Dana	*	..	..	..
27	,, ,, cyclas, J. D. Dana	*	..	..	..
28	,, ,, transversus, J. D. Dana	*	..	..	..
29	Pachydomus ? corpulentus, J. D. Dana	*	..	..	..
30	Pachydomus ,, intrepidus, J. D. Dana	*	..	..	..
31	,, ,, lineatus, J. D. Sowerby	*	..	..	..
32	,, ,, antiquatus, J. D. Sowerby	*	..	..	..
33	,, ,, sacculus, F. McCoy	*	..	..	..
34	,, ,, lævis, J. D. Sowerby	*	..	..	..
35	Cardinia ? recta, J. D. Dana	*	..	..	..
36	Eurydesma cordata, J. Morris	*	..	..	..
37	Eurydesma ? elliptica, J. D. Dana	*	..	..	..
38	Eurydesma ? globosa, J. D. Dana	*	..	..	..
39	Cypricardia ? acutifrons, J. D. Dana	*	..	..	..
40	Cypricardia ? imbricata, J. D. Dana	*	..	..	..
41	Cypricardia ? arcodes, J. D. Dana	*	..	..	..
42	Cypricardia ? prærupta, J. D. Dana	*	..	..	..
43	Cypricardia ? siliqua, J. D. Dana	*	..	..	..
44	Cypricardia ? simplex, J. D. Dana	*	..	..	..
45	Cypricardia (Avicula?) Veneris, J. D. Dana	*	..	..	..
46	Venus ? gregaria, J. D. Dana	*	..	..	..
47	Notomya ? securiformis, F. McCoy	*	..	..	..
48	Notomya ? clavata, F. McCoy	*	*	..	..
49	Orthonota ? compressa, J. Morris	*	..	..	..
50	Orthonota ? costata, J. Morris	*	..	..	..
51	Mæonia valida, J. D. Dana	*	..	..	..
52	,, axinia, J. D. Dana	*	..	..	..
53	Mæonia ? carinata, J. D. Dana	*	..	..	..
54	Mæonia fragilis, J. D. Dana	*	..	..	..
55	,, rugiformis, J. D. Dana	*	..	..	..
56	,, elliptica, J. D. Dana	*	..	..	..
57	,, grandis, J. D. Dana	*	..	..	..
58	Mæonia ? recta, J. D. Dana	*	..	..	..
59	Tellinomya (Nucula) abrupta, J. D. Dana	*	..	..	..
60	,, ,, concinna, J. D. Dana	*	..	..	..
61	,, ,, Glendonensis, J. D. Dana	*	..	..	..
62	Pinna ? (Carilium) ferox, J. D. Dana	*	..	..	..
63	Modiola crassissima, F. McCoy	*	*	..	..
64	Aviculopecten squamuliferus, J. Morris	*	..	..	..
65	,, comptus, J. D. Dana	*	..	..	..
66	,, tenuicollis, J. D. Dana	*	..	..	..
67	,, mitis, J. D. Dana	*	..	..	..
68	,, imbricatus, R. Etheridge	..	..	*	..
69	,, (Streptorynchus) Davidsoni, R. Eth.	..	..	..	..
70	Avicula Volgensis ? E. de Verneuil	..	..	..	*
71	Theca lanceolata, J. Morris	*	..	..	..
72	Conularia ? torta, F. McCoy	*	..	..	..
73	Pleurotomaria nuda, J. D. Dana	..	..	..	..
74	,, Strzeleckiana, J. Morris	*	..	*	*
75	,, carinata, J. Sowerby	..	..	..	*
76	Bellerophon decussatus, Fleming	*	..	..	..
77	Euomphalus depressus, J. D. Dana	*	..	..	..
78	,, (Platyschisma) rotundatus, J. Morris	*	..	..	..
79	Naticopsis ? harpæformis, R. Etheridge	..	..	*	..
80	Bairdia affinis, J. Morris	*	..	..	..
81	,, curta, F. McCoy	*	..	..	..
82	Cythere impressa, F. McCoy	*	..	..	..
83	Urosthenes Australis, J. D. Dana	*	..	..	..
	TOTAUX	73	5	8	7
	En y ajoutant les totaux du tableau précédent	176	9	12	74
	on aura pour totaux généraux	249	14	20	81

dont cent une, ou les deux cinquièmes à une petite fraction près [? cent soixante ou cinquante-neuf ou presque les *trois* cinquièmes.—W.B.C.] se trouvent exclusivement dans la Nouvelle-Galles du Sud et n'ont jusqu'ici de représentants dans aucun autre pays.

Il est à remarquer qu'un petit nombre de ces espèces appartiennent à des genres qui n'existent pas en Europe. Telles sont : les *Tribrachiocrinus*, les *Clarkea*, les *Eurydesma*, les *Aphanaia* et les *Urosthenes*.

En jetant un coup d'œil sur les planches qui accompagnent mon travail, on pourra se convaincre, en outre, que plusieurs espèces ont pris un développement extraordinaire. Je citerai, entre autres, le *Cyathocrinus* (*Konincki*, W. B. Clarke, les *Spirifer glaber*, W. Martin, *Darwinii*, J. Morris, quelques espèces de *Pachydomus* et de *Mæonia* l'*Aphanaia gigantea*, L.-G. de Koninck, les *Aviculopecten Illawarrensis* et *limæformis*, J. Morris, et les *Conularia inornata*, J. D. Dana.

On serait tenté de croire que ces espèces ont été soumises à des influences spéciales ayant favorisé leur croissance, si, à côté d'elles, il ne s'en trouvait d'autres, qui n'atteignent pas la moitié de la taille qu'elles possèdent généralement en Europe. Telles sont les *Loxonema constricta*, W. Martin, le *Macrocheilus acutus*, Sowerby, et la plupart des Gastéropodes.

Afin de déduire de l'ensemble des espèces décrites, la stratification des terrains qui les ont fournies, j'ai dû me borner à faire usage des quatre-vingt-une espèces européennes que l'on compte parmi elles et de rechercher les assises dans lesquelles elles ont été découvertes.

Cet examen m'a fourni la preuve que vingt-deux de ces espèces étaient communes aux assises tant supérieures que moyennes et inférieures du calcaire Carbonifère, que trente-six appartiennent exclusivement aux assises supérieures, cinq ou six à la fois aux assises supérieures et moyennes et enfin six ou sept aux assises inférieures. Mais il est à remarquer que tandis que les trente-six espèces supérieures renferment un certain nombre d'espèces caractéristiques, telles que les *Lithostrotion basaltiforme* et *irregulare*, les *Productus fimbriatus*, *punctatus* et *undatus*, le *Chonetes papilionacea*, le *Spirifer bisulcatus*, les *Pleurotomaria gemmulifera* et *carinata*, l'*Enomphalus catillus*, le *Loxonema constricta*, etc., les assises moyennes et inférieures ne fournissent aucune des espèces qui les font facilement reconnaître ; telles sont, entre autres, pour les premières le *Spirifer striatus* et le *Syringothyris cuspidatus* et pour les secondes l'*Athyris Royssii*, les *Spirifer Mosquensis* et *laminosus*, le *Conocardium Hibernicum* et le *Nautilus Konincki* qui y font complétement défaut.

Je crois donc être en droit de conclure que la plupart des roches Carbonifères de la Nouvelle-Galles du Sud appartiennent aux assises supérieures du terrain ; qu'une partie, principalement celle qui renferme les *Spirifer convolutus et pinguis*, var. *rotundatus*, peut être attribuée aux assises moyennes, et que si les assises inférieures y sont représentées ce n'est que par quelques lambeaux insignifiants ou du moins très-pauvres en fossiles. Je laisserai à d'autres les déductions biologiques que l'on pourra tirer de l'étude de la faune Carbonifère que je viens de décrire et de la comparaison avec celle des autres pays.

Je me bornerai à faire remarquer qu'il est probable que la mer dans laquelle se sont développés les animaux Carbonifères de l'Australie, était en communication avec celle dans laquelle ont vécu les animaux de la même époque qui se trouvent actuellement en Belgique, aux environs de Visé et de Namur; en Angleterre dans le Yorkshire; en Ecosse aux environs de Glasgow ; en Irlande près de Cork et de Dublin, et en Allemagne dans la Silésie. Cette mer existait encore alors que déjà la majeure partie des roches Carbonifères de l'Amérique et de la Russie, ainsi que celles du Nord de l'Irlande et des environs de Tournai, de Feluy, de Soignies et de Comblain-au-Pont de notre pays, étaient déjà émergées et que les animaux qu'elles renferment étaient en majeure partie détruits.

APPENDIX XVII.

Mr. Lonsdale's List, in 1858, of New South Wales Zoantharia submitted to him for examination by Rev. W. B. Clarke in 1855.

Genera.	No. of Species.	Number of Species in Europe and America belonging to—			
		Lower Silurian.	Upper Silurian.	Devonian.	Carboniferous.
Favosites	2	4	4	3	......
Calamopora	1	2	3 (?)	4	......
Emmonsia	2	......	1	1	1
Alveolites	3 (?)	2	4	5	......
Syringopora	3	1	4	5	6
New genus (?)	1	......	......	......	......
Cœnites	1	......	4	2	......
Cladopora (?)	1	......	7	......	......
Endophyllum	1	......	......	2	......
Trochophyllum	1	......	......	......	1
Ptychophyllum	1	1	1	1	......
Clisiophyllum	1	......	2	......	6
Cystophyllum	1	1	3	3	......
Heterophyllia	1	......	......	......	2
	20	11	33	26	16

Extract from letter from Mr. Lonsdale in explanation of the foregoing List:—

"My dear Sir,   "Bradford-on-Avon, Wilts, 12 July, 1858.

"I have not been able to reply sooner to your letter of 13th March, and the answer which I now give will prove I fear very unsatisfactory. The accompanying table contains a list of genera, and the number of species; but the species are believed to be all new, and three or four of the genera are doubtful determinations. It is impossible, therefore, to define precisely the position of the beds which afforded the corals; but the table contains also a rough enumeration of the geological distribution in Europe and America of the Australian genera; with some of the known localities. These details may assist in approximating towards the age of the beds which yield your Zoantharia. The genera Favosites, Calamopora (as distinct from Favosites), Emmonsia, Alveolites, Cœnites, Ptychophyllum, and Cystophyllum may be assumed to be essentially Upper Silurian or Devonian, but it is impossible to state whether the Australian strata belong to the former or to the latter or to both. Negative evidence based on the absence of Halysites would lead to the conclusion that the deposits are Devonian. Such evidence is, however, valueless, as an additional number of specimens might afford evidence of the existence of Halysites and other important genera. As respects the Carboniferous series, the genera Syringopora and Clisiophyllum have each six species, principally from British localities; but Syringopora has four or five Silurian and five Devonian, and these (4 + 5) neutralize any deduction which might be drawn from the six Carboniferous. Clisiophyllum has two Silurian representatives, though as far as known no Devonian. The genus might consequently be deemed to indicate a Carboniferous age; it cannot, however, be opposed to seven genera (Favosites, &c.) above mentioned. Favosites has been said to occur in the mountain limestone, but I have not seen an example of it, and the genera as generally received requires much additional consideration.

"Rev. W. B. Clarke, &c., &c.,   "Ever your most obliged,
"New South Wales."   WM. LONSDALE.

COLLECTION OF FOSSILS made by Rev. W. B. Clarke, and forwarded to the Woodwardian Museum, Cambridge; borrowed from Prof. Sedgwick by Sir R. I. Murchison, in 1856, for examination and description by Mr. Salter, from whose MS. notes, sent to Mr. Clarke in 1858, the following are named, in addition to a series of *Zoantharia* examined by the late W. Lonsdale, Esq., F.G.S. [See "*Remarks on Sedimentary Formations of New South Wales, by Rev. W. B. Clarke,*" ("*Mining and Mineral Statistics,*" 1873, p. 157); and Murchison's "*Siluria,*" Third Ed., p. 296, and Fourth Ed., p. 276 and p. 462.]

Genus.	Species.	Locality.
	UPPER SILURIAN.	
Alveolites	(?) oculata	
= Millepora	repens.	
Cyathophyllum		
Favosites	polymorpha	Snowy River basin.
Heliolites		Maneero.
Syringopora	n. sp.	Limestone Creek.
Tentaculites	very like "ornatus."	
Beyrichia		
Calymene	? Blumenbachii	
„	Macleayi	Yarralumla.
Encrinurus	Australis	id.
Pentamerus	do.	id.
„	n. sp. "plaited"	Coolalamine; Quedong, &c.
Euomphalus	alatus.	
Trochonema		Yarradong.
Bellerophon		
Orthoceras		id.
Receptaculites	Clarkei (MS.)=Australis. ["*Geol. Survey of Canada,*" Decade I—Organic Remains, p. 47, pl. x, fig. 8-10.]	id. [and Quedong.]
	DEVONIAN.	
Atrypa	reticularis.	
Orthis	resupinata.	
Productus, or Chonetes.		
Spirifer	five species.	
Strophomena		
Euomphalus		
Loxonema	n.s. allied to "*rugifera.*"	
Murchisonia	like *angulata.*	
Pluncrotinus		

P.S.—"Except perhaps Atrypa *reticularis,* Favosites *polymorpha,* Alveolites *oculata,* I do not recognize any undoubted British species.— J. W. SALTER."

Remarks on the preceding Lists.

The arrangement of the above fossils cannot be considered entirely satisfactory, and it is due to the memory of my lately departed friend Mr. Salter to give some explanation of the matter. It had long been my desire to place in the Woodwardian Museum, at Cambridge, as nearly a complete series of rocks and fossils from New South Wales as I could obtain in the course of my explorations in the Colony.

In November, 1844, I forwarded to my friend Prof. Sedgwick four large casks containing the first series from districts named in Appendix III. One of these casks, it appears, did not reach its destination, and must have been lost on the voyage, or on its way to Cambridge. From the remainder Prof. M'Coy, then engaged with the Woodwardian Palæozoic collections, afterwards so ably discussed and described, in 1855, in the joint volume of the two geologists ("*Synopsis of the Classification of the British Palæozoic Rocks, with a systematic description of the British Palæozoic Fossils in the Geol. Museum of the Univ. of Cambidge*"), undertook to describe and publish at his learned colleague's request and charge, in the "*Annals & Magazine of Nat. History*, vol. xx," many of the vegetable and Marine fossils that remained of my collection, under the title of—"*On the Fossil Botany and Zoology of the Rocks associated with the Coal of Australia,*" and in one of the present lists these so-described fossils have been enumerated.

In the year 1855, having entered upon a new field of research—the former having been confined chiefly to the examination of the Coal Measures and the Marine Fossils of the Upper Palæozoic associated with them, as will be seen by the letter (c) in the list already referred to—and having obtained a considerable collection of fossils from the Middle and Lower Palæozoic rocks, I forwarded to my friend Professor Sedgwick, in continuation of my purpose of completing the exhibition of the New South Wales fossil succession, a series of such fossils as would show that below the Carboniferous strata, Devonian and at least Upper Silurian formations exist in this Colony.

No description of the fossils having been obtainable from Cambridge, I wrote both to Prof. Sedgwick and to Sir R. I. Murchison, the latter of whom borrowed them at my request from the former, and submitted them to Messrs. Lonsdale and Salter, who did their best to meet the necessity, but could not complete the work.

A letter of the former I have already put in evidence, and extracts of letters from the latter will be appended to these remarks, which are made public in justice to my friends Sedgwick, Murchison, Lonsdale, and Salter, all of whom are now deceased.

Subsequently to this I entered into further arrangement with Mr. Salter, who undertook to complete a description, with figures, of a considerable number of Lower Palæozoic and Devonian species, but his death prevented the work. This collection, therefore, was left undescribed, except in the way recorded, till after the deaths of Murchison and Sedgwick. Not being able to know what had been done with them by the former, I wrote to Professor Hughes, his successor at Cambridge, who very promptly informed me that on inquiry he was unable to learn what had become of these fossils, or whether they had been returned to the Cambridge Museum—which, of course, he could not determine from personal knowledge. As Mr. Salter said, that with those he named there were other "several beautiful species," it is possible that some valuable additions to palæontology may have occurred, as he made particular request respecting certain of them in the following memorandum :—"We should like to have the localities of the above numbers, and of the following corals and shells, viz. : —

    Massive Favosites, 2,983, 3,526, 3,540.
    Cylindrical one, 2,507.
    Flat one, 3,602.
    Syringopora —large, 3,540.
    New genus, allied to Favosites, 3,616.
    New genus, 3,597.
    New genera, 3,553, 3,562, 3,588.
    Heliolites, 3618."

The localities were supplied by me.

In the same Memo. he says :—"Mr. Lonsdale is examining a few of the corals. If he should be able to throw any light on which is Devonian, &c., I am sure Sir Roderick will send you the information." This was done, so far as was possible, as I find several references to them, and to the endeavours of Messrs. Salter and Lonsdale, in various letters from Sir Roderick, who, independently of his private correspondence, made public mention of them in "*Siluria*," 4th Ed., pp. 18, 462, 276. Murchison further says, in one of his letters to me, that had I sent my fossils to what he considered their "proper destiny"—the School of Mines—they would have been officially described long before.

Writing on November 16, 1853, he says :—"I am always glad to receive your instructive letters, and was much pleased to find that you have been throwing so much important light on the auriferous phenomena of Australia. I will not fail to profit by your discoveries in the golden chapter of my forthcoming "*Siluria*."…. I have long been anxious of having your Palæozoic fossils properly named and compared before my final chapter is printed."

The above extracts are here introduced to show that no possible purposed neglect occasioned the disappointment as to my earlier description of the Palæozoic collections made and forwarded by me to Europe. I may add further, that acting on a suggestion of Mr. Salter that I should apply to Prof. M'Coy, "who is well qualified," I did so, feeling that as I had been indebted to him for the description of the Carboniferous Fossils from the first contribution to Cambridge, it would have been gratifying to me to have placed in his hands the Middle Lower Palæozoics of the second contribution.

The learned Professor, in reply, stated that his engagements of a public character were too onerous to allow him time to devote to more private work of the kind, and courteously declined.

In this extremity I consulted Prof. T. Rupert Jones, who recommended me to seek aid from Prof. De Koninck, of Liége, who—after some delay on my part, occasioned by circumstances which did not originate in any want of continued zeal, but over which I had no control—has most ably, indefatigably, and willingly accomplished it, to his own reputation as I hope and believe, and certainly with much honorable acknowledgment of myself.

I had myself begun the work in a small way by making drawings to scale of more than 1,200 individual specimens collected by me from the Carboniferous beds, chiefly between the years 1843 and 1847, and including many described by Prof. M'Coy and De Koninck, and which were shown to the former in the year 1860.

They were never published, and it was the feeling that it was a work beyond my own powers to do justice to it, coupled with want of pecuniary means and of leisure from my parochial duties, that induced me to look to professional and acknowledged authorities in Palæontology out of the Colony, and I am proud to acknowledge that I have found many able and willing condjutors who have in many instances given me gratuitously the heartiest and most disinterested assistance.

It is with the intention of acknowledging this aid that I have said so much respecting my two friends—Lonsdale, with whom I became acquainted more than half a century ago, and Salter, with whom my acquaintance was more recent.

The following are extracts from the letters of Mr. Salter in continuation of former quotation:—

"Museum of Pract. Geology,
"Jermyn-street, London,
"May 9, 1856.

"I should have answered your letter some time ago (I have had it a few weeks only in hand), but for the very reason that prevents my being able to work at the Australian fossils in the way I should like. I thought to have given you some additional information respecting them, but it is, I find, impossible at present, owing to the pressure of work falling on my department—a sad hiatus being made in all my calculations, and a period put to much important work by the lamented death of Forbes. He could scarcely overtake the work that necessarily falls on those who have to help every one with fresh studies in the fossil groups, and how am I to expect to do it? We are finishing off our own Silurian work for England, and besides are compelled to attend to the wants of all the other departments of British geology. Under these circumstances, it will be utterly impossible to make any fresh detailed examination of the fossils you mention. The abstract sent to you by Sir Roderick will have clearly answered one of your most important queries, since there can be no doubt of a true Upper Silurian formation among your fresh fossils; the presence of *Calymene, Encrinurus*, and a plaited *Pentamerus* quite settles that question. *Receptaculites*, too, is a good Silurian genus when combined with such fossils as the above.

"I have some time had in contemplation to give a short paper on some fossils of Yarralumla, collected long ago, which are undoubtedly Silurian, but now your new fossils have arrived it will enable me to add a figure or two of the principal species of these, when a temporary leisure may enable me to attack them. It will not be at present, but it will be a pleasure to me when I have the time. Yours very sincerely,
J. W. SALTER."

"Mus. Pract. Geology, Jermyn-st.,
"London, Nov., 28, 1858.

"Your letter, just come to hand, convinces me that your Colony will no longer need any illustration from Home.

"The specimens which you kindly sent to Cambridge, and which I have examined in the rough, are, I am glad to find, only duplicates. I have not attempted to define their species. To do so would be to work them out, and I could, of course then send you an account of them easily. My avocations do not permit me any leisure; and though the presence of the genera I mentioned do certainly indicate the Upper Silurian, yet the great abundance of corals, both millepore and cupcorals, with *Productus* and *Atrypa reticularis*, is an association we never meet with here below Devonian, still I cannot give you any specific names, except the very few opposite (vide list p. 151). The Receptaculites is a beautiful thing that helps to illustrate the genus for which I have long ago had materials, and if you allow me to describe that species along with another from Canada I shall be glad to do so.

"Yours faithfully,
"J. W. SALTER."

EXTRACT from "*Canadian Fossils*," decade x, p. 47.

"R. *Australis* n. sp.   Plate x.   Fig. 8–10.

"Spec. character.—R. magnus, expansus, cellulis verticalibus, subcylindricis, incrassatis, apicibus subter convexis, tabulatis.

"Under this name a curious species of the genus is figured, for the sake of comparison, from the Silurian limestones of New South Wales, communicated by the Rev. W. B. Clarke.

"It is remarkable as having the expanded apices of the columns on the lower surface, tabulated in larger or smaller divisions which all seem to radiate from a central boss. And this arrangement is quite different from the merely granulated surface observable in the *R. occidentalis* (formerly *R. Neptuni*.) Locality, Upper Silurian limestone of Yarradong, between the Yass Plains and the Murrumbidgee River, New South Wales—a locality rich in Upper Silurian Forms, Tentaculites, Favosites, Pentamerus, Orthoceras, Trochoceras, Rhynconella, &c.

"Feb., 28th, 1859.   J. W. SALTER."

APPENDIX XVIII.

SCHEMES OF ARRANGEMENT, by different authors, of the PALÆOZOIC FOSSILS of the N. S. W. SEDIMENTARY FORMATIONS.

	Genus.	Species.	Authority.	W.B.C.	M'Coy.		Feistmantel.	Dana.
Hawkesbury and Wianamatta	Palæoniscus	antipodeus	Eg.					
	Myriolepis	Clarkei	Eg.					
	Cleithrolepis	granulatus	Eg.		Supra-carboniferous	Oolite	Rhætic and Keuper Upper Trias.	Permian in part.
	Sphenopteris	alata	Mor.					
	Pecopteris = (Thinnfeldia)	odontopteroides	Feist.					
	Do.	tenuifolia	M'C.					
	Odontopteris	microphylla	M'C.					
	Phyllotheca	Hookeri	M'C.					
	Gleichenia	sp.	M'C.					
	Echinostrobus	sp.	Feist.					
	Tæniopteris	Wianamatta	Feist.					

No Marine Fossils.

	Genus.	Species.	Authority.
Newcastle, Bowenfells, Lithgow, Illawarra.	Urosthenes	Australis	Dan.
	Glossopteris	Browniana.	
	Do.	reticulum	Dan.
	Do.	elongata	Dan.
	Phyllotheca	Australis.	
	Do.	ramosa.	Dan.
	Noeggerathia	spatulata	Dan.
	Do.	media	Dan.
	Do.	elongata	Mor.
	Zeugophyllites	sp.	Do.
	Vertebraria	sp.	
	? Otopteris	ovata	M'C.
	Sphenopteris	lobifolia	Mor.
	Do.	alata.	
	Gangamopteris	angustifolia	M'C.
	Tæniopteris	sp.	
	Brachyphyllum	Australe	Feist.

Marine Carboniferous beds.
Stony Ck., Greta, Anvil Ck., Rix's Ck., Mt. Wingen, &c.

APPENDIX XVIII—continued.

	Genus.	Species.	Authority.
Carboniferous.	Glossopteris	Browniana	Feist.
	Do.	Do. præcursor	Do.
	Do.	Clarkei	Do.
	Do.	primæva	Do.
	Macrotæniopteris	sp.	Do.
	? Lepidodendron	Australe	M'C.

Marine Carboniferous animals.

	Genus.	Species.	Authority.
Icthyodorulite Ranges, Port Stephens.	Lepidodendron	nothum.	
	Do.	Veltheimianum.	
	Cyclostigma?	Kiltorkauense.	
	Do.	intermedia.	
	Glossopteris	primæva	Feist.
	Sigillaria	Bornia.	
	Schizopteris	radiata.	

APPENDIX XIX.

ALTHOUGH no Mesozoic Marine Fossils have been discovered in New South Wales, it has been thought desirable to publish the following Lists of Fossils of this formation discovered in the other Colonies. They are extracted from the "*Quarterly Journal Geological Society*" for 1870, pp. 231, 232, 239, 240, from a paper on "*Australian Mesozoic Geology and Palæontology, &c.,*" by *Chas. Moore, Esq., F.G.S.*

General List of Organic Remains from Western Australia.

Plantæ.
Cliona (?).
Cristellaria cultrata, *Montf.* * (O.)
Echini (spines).
Serpulæ.
Entomostraca, sp.
Polyzoa, sp.
Rhynchonella variabilis, *Schloth.* * (O.)
Avicula Münsteri. *Goldf.* * (O.)
—— echinata, *Sow.* * (O.)
—— inæquivalvis, *Sow.*

Lima proboscidea, *Sow.* * (O.)
—— punctata *Sow.* * (O.)
—— duplicata, *Sow.*
—— sp.
Lima sp.
Ostrea Marshii, *Sow.* * (O.)
Ostrea two sp.
Plicatula sp.
Pecten cinctus, *Sow.* * (O.)
—— calvus, *Münst.* * (O.)
—— Greenoughiensis, *Moore.*
Astarte Cliftoni, *Moore.*

APPENDIX XIX—*continued.*

Astarte apicalis, *Moore.*
—— two sp.
Cardium, sp.
Cucullœa oblonga, *Sow.* * O.)
—— three sp.
Cypricardia, sp.
Gresslya donaciformis, *Ag.* * (U.L.)
Isocardi, sp.
Myacites liassianus *Quenst.* * (M.L.)
—— Sanfordii, *Moore.*
—— two sp.
Pholadomya ovulum, *Ag.* * (O.)
Teredo Australis, *Moore.*
Tancredia, sp.
Trigonia Moorei, *Lycett.*
Unicardium, sp.
—— (?), sp.
Amberleya, sp.

Cerithium, sp.
Eulima (?), sp.
Phasianella, sp.
Trochus, sp.
Turbo lævigatus, *Sow.*
—— sp.
Rissoina Australis, *Moore.*
Ammonites nalensis, var. Moorei,
    Lycett * (U.L.)
—— radians, *Rein.* ° (U.L.)
—— Brocchii. *Sow.* * (O.)
—— macrocephalus, *Schloth.* * (O.)
—— Walcottii, *Sow.* * (U.L.)
——, sp.
Nautilus semistriatus, *d'Orb.**(U.L.)
Belemnites canaliculatus, *Schloth.*
    * (O.)

NOTE.—In the above list the * marks species common to Western Australia and England, and the formations in which they occur in the latter country are placed within (), "O." meaning *Oolite*, "L." *Lias*, ("M." *Middle*, "U." *Upper*.)

APPENDIX XIX (2.)

List of Mesozoic Species from Queensland.

Plantæ (wood).
Purisiphonia Clarkei, *Bowerbank.*
Cristellaria acutauricularis, *Ficht. &
    Moll.*
—— cultrata, var. radiata, *Moore.*
—— acutauricularis, var. longicostata.
    Moore.
Dentalina communis, *d'Orb*
Balanus, sp.
Entomostraca, sp.
Lepralia oolitica, *Moore.*
Polyzoa, sp.
Argyope Wollumbillaensis, *Moore.*
—— punctata, *Moore.*
Discina apicalis, *Moore.*
Lingula ovalis, *Sow.*
Rhynchonella rustica, *Moore*
—— solitaria, *Moore.*

Terebratella Davidsonii, *Moore.*
Avicula simplex, *Moore.*
—— æqualis, *Moore.*
—— Braamburiensis, *Phil.*
—— Barklyi, *Moore.*
—— substriata, *Moore.*
—— reflecta, *Moore.*
—— umbonalis, *Moore.*
—— corbiensis, *Moore.*
—— sp.
Lima Gordonii, *Moore.*
—— sp.
—— sp.
—— multistriata, *Moore.*
Pecten æquilineatus, *Moore*
—— socialis, *Moore.*
—— fimbriatus, *Moore*
—— sp.

APPENDIX XIX (2)—continued.

Pecten sp.
—— sp.
Perna gigantea, *Moore*.
Arca plicata, *Moore*.
—— prælonga, *Moore*.
Astarte Wollumbillacusis, *Moore*.
Cardinia, sp.
Cardium, sp.
Cytherea Clarkei, *Moore*.
—— gibbosa, *Moore*.
Polymorphina lactea, *W. & J.*
—— gibba (?), *d'Orb.*
Planorbulina Ungeriana, *d'Orb.*
—— lobatula, *d'Orb.*
Vaginulina striata, *d'Orb.*
Pentacrinus Australis, *Moore*.
Echinus (spines).
Serpula intestinalis, *Phil.*
Goniomya depressa, *Moore*.
Leda Australis, *Moore*.
Lucina anomala, *Moore*.
—— Australis, *Moore*.
Mactra trigonalis, *Moore*.
—— sp.
Modiola unica, *Moore.*
Mya Maccoyi, *Moore.*
Myacites planus, *Moore*.
Mytilus inflatus, *Moore*.

Mytilus rugo-costatus, *Moore*.
—— planus, *Moore*.
Nucula, Cooperi, *Moore*.
—— truncata, *Moore*.
—— sp.
Panopæa rugosa, *Moore*.
Taucredia plana, *Moore*.
Thracia Wilsoni, *Moore*.
Trigonia lineata, *Moore*.
Actæon Hochstetteri, *Moore*.
—— depressus, *Moore*.
Delphinula reflecta, *Moore*.
Dentalium lineatum, *Moore*.
Natica variabilis, *Moore*.
—— ornatissima, *Moore*.
—— sp.
Solarium (?), sp.
Trochus, sp.
Turbo, sp.
Belemnites paxillosus (?), *Voltz.*
—— Australis, *Phillips.*
—— sp.
—— sp.
Crioceras Australe, *Moore* [*Neocomian.*]
Teuthis, sp.
Hybodus ? (teeth and scales).
Lepidotus (scales).

General Table of Secondary Species.

	No. of species.		No. of species.
Plantæ	2		23
Amorphozoa	2	Brachiopoda	8
Foraminifera	7	Conchifera	83
Echinodermata	4	Gasteropoda	18
Articulata	4	Cephalopoda	13
Crustacea (Entomostraca)	2	Pisces	3
Polyzoa	2		
	23		148

[Reptilian remains have been discovered at the Flinders River—Cretaceous.]

See also Paper by Rev. W. B. Clarke—"*On Marine Fossiliferous Secondary Formations in Australia*": Q. J. G. S., vol xxiii (1867), p. 7.

APPENDIX XX.

CORRELATION of Australian Fossils, exclusive of Marine, by Ottokar Feistmantel, M.D., Palæontologist to the Indian Geological Survey, in manuscript letter of 26th February, 1878.

Systematical Table.

NOTE.—The species marked "n. sp." are described by him as new.

Name.	Strata above Marine fauna.				Marine beds with Palæozoic fauna.			
	Lyremost (Jurassic) beds ("Trenlopteris beds(?), Tasmania (?), Victoria, & Clarence R.	Wianamatta and Hawkesbury.	(?) Bacchus Marsh Sandstones (Lower Mesozoic in Victoria), Newcastle beds =	Upper Coal Measures in N.S.W.	Lower Coal Measures, Mesozoic flora with Palæozoic fauna.	Strata with Lower Carboniferous flora, Smith's a Creek; Port Stephens.	Upper Devonian (?).	
ANIMALS (Fishes).								
Urosthenes Australis, Dan.	...	...	...	*	...	...	...	a. Strata with plants which by themselves are of *Carboniferous* (partly Lower Carb.) age.
Palæoniscus antipodeus, Eg.	...	*	...	...	...	...	...	b. Strata the plant remains of which are not by themselves of Carboniferous age, but have to be considered of that age on account of association with Marine fossils of Carboniferous age, as Anvil Creek, Ilk's Creek, &c.
Clethrolepis granulatus, Eg.	...	*	...	...	...	...	...	
Myriolepis Clarkei, Eg.	...	...	...	...	...	...	...	c. See note at end.
PLANTS.								
1. Equisetaceæ.								
Phyllotheca Australis, Bgt. (including two other species.)	*	...	...	*	*	...	...	Heteroveral. Newcastle. Heteroveral. Wianamatta. Not heterov. [but Egerton distinctly calls it heterov.—W.B.C.] Tail not known. Hawkesbury. Anvil Creek, quoted by W. B. C. and C. H. Wilkinson; Newcastle and Clarke's Hill (Cobbitee) Wianamatta; [Ph. *Hookeri*, M'Coy.] Victoria, M'Coy.
Vertebraria Australis, M'Coy (including Glossteria of Dana.)	...	...	...	*	...	...	...	Newcastle; Blackman's Swamp.
Sphenophyllum, sp.	...	...	...	...	...	*	...	Port Stephens. Lowest Carboniferous. (?) Ursa *stufe* of Ileer.
II. Filices.								
1. *Sphenopterides.*								
Sphen. lobifolia, Morr.	...	*	...	*	...	*	...	Newcastle. Hawkesbury. Is not a Palæozoic form of Germany.
,, alata, Bgt.	...	...	...	...	...	...	...	Newcastle.
,, ,, var. exilis, Morr.	...	...	...	...	...	...	...	

160 *Sedimentary Formations*

APPENDIX XX—*continued.*

Column headers (Strata above Marine fauna):
- Uppermost (Jurassic) beds ("Tæniopteris beds"), Tasmania (?), Victoria, & Clarence R.
- Wianamatta and Hawkesbury.
- (?) Bacchus Marsh Sandstones (Lower Mesozoic in Victoria).
- Newcastle beds = Upper Coal Measures in N.S.W. (c)
- Lower Coal Measures, Mesozoic flora with Palæozoic fauna. (b)
- Strata with Lower Carboniferous flora, Smith's Creek; Port Stephens. (a)
- Upper Devonian (?)

Marine beds with Palæozoic fauna:
a. Strata with plants which by themselves are of Carboniferous (partly Lower Carb.) age.
b. Strata the plant remains of which are not by themselves of Carboniferous age, but have to be considered of that age on account of association with Marine fossils of Carboniferous age, as Anvil Creek, Rix's Creek, Greta, &c.
c. See note at end.

Name.	Uppermost (Jurassic)	Wianamatta	Bacchus Marsh	Newcastle / Upper Coal	Lower Coal Measures	Strata w/ Lower Carb.	Upper Devonian (?)	Locality
Sphen. hastata, M'Coy				*				Newcastle.
,, elongata, Carr.				*				Tivoli Mines, Queensland.
,, Germana, M'Coy				*				Newcastle.
,, plumosa ,,				*				Newcastle.
,, flexuosa ,,				*		*		Smith's Creek .... } Lowest Carboniferous. Smith's Ck.; Port Stephens } (?) *Ursa Stufe* of Heer.
Rhacopteris intermedia, n. sp.						*		
,, inaequilatera, Göpp.						*		
2. *Neuropterides.*								
Thinnfeldia odontopteroides, Morr, s.p.	*	*		*	*?			b. (?) Jerusalem Basin, Tasmania; Newcastle; Clarke's Hill (Wianamatta); Ipswich, Tivoli Mines, Queensland; Victoria.
Odontopteris microphylla, M'Coy				*				Newcastle: Clarke's Hill (Wianamatta).
,, [? (?)opteris] ovata, M'Coy, sp.				*?				(?) Arrow [Paterson R., believed to be an error, stated to be *Rhacopteris*, Feist. W. k C.]
Cyclopteris cuneata, Carr.				*				Tivoli Mines, Queensland.
3. *Pecopterides.*								
Alethopteris Australis, Morr.	*?	*		*	*?			(?) Jerusalem Basin, Tasmania; Cape Paterson and Leeherine, Victoria; Clarence River, N.S.W.; Tasmania.
Pecop. (?) tenuifolia, M'Coy		*		*				Newcastle: Clarke's Hill (Wianamatta).
Gleichenia dubia, n. sp.		*		*				Clarke's Hill.
4. *Tæniopterides.*								
Tæniopt. (Angiopteridium) Daintreei, M'Coy	*	*						Cape Paterson; Victoria; Queensland; Clarence River, N.S.W. W. B. C.'s specimens.
,, (Macrotæniopt.) Wianamattæ, n. sp.		*?						Wianamatta, Gib Tunnel, Merrigang.

APPENDIX XX—continued.

New South Wales.

Name.	Uppermost (Jurassic) beds ("Tæniopteris beds"), Tasmania (?), Victoria, & Clarence R.	Wianamatta and Hawkesbury.	(?) Bacchus Marsh Mesozoic Sandstones (Lower in Victoria).	c. Upper Coal Measures = Newcastle beds in N.S.W	b. Lower Coal Measures, Mesozoic flora with Palæozoic fauna.	a. Strata with Lower Carboniferous flora, Smith's Creek; Port Stephens.	Upper Devonian (?)	Marine beds with Palæozoic fauna.
								a. Strata with plants which by themselves are of *Carboniferous* (partly Lower Carb.) age.
								b. Strata the plant remains of which are not by themselves of Carboniferous age, but have to be considered of that age on account of association with Marine fossils of Carboniferous age, as Anvil Creek, Rix's Creek, Greta, &c.
								c. See note at end.
5. *Dictyopterideæ*.								
Glossopteris Browniana—								
" var. pracursor, Bgt.	* ?							Stony Creek, N.S.W.
" var. genuina, Bgt.	*							Tasmania (?) ⎫ First appearance of
" mentioned generically								N.S.W. and Queensland ⎬ Glossopteris
" primæva, n. sp.								Greta, N.S.W. ⎭ below Marine beds.
" Clarkei, n. sp.								Rix's Creek, N.S.W.
" Browniana								Newcastle; Illawarra
								Blackman's Swamp
" linearis, M'Coy					*			Arowa; Newcastle; Wollongong
" ampla, Dan.				* * *				Newcastle ⎫ Second appearance of
" reticulum, Dan.				*				Newcastle ⎬ Glossopteris above
" elongata, Dan.				* * * * * * * * *				Newcastle ⎭ Marine beds.
" cordata, Dan.								Newcastle ; Illawarra
" Temlopteroides, n. sp.			*					Blackman's Swamp ;
" Wilkinsoni, n. sp.			* *					Newcastle ....
" parallela, n. sp.								
" young leaves								Newcastle; Gantawang (Mudgee); Bacchus Marsh.
Gangamopteris angustifolia, M'Coy								Newcastle; Blackman's Swamp.
" Clarkeana, n. sp.								Bacchus Marsh.
" spathulata, M'Coy						*		Bacchus Marsh.
" obliqua, M'Coy								Tasmania, W. B. C.'s specimens.
Sagenopteris Tasmanica, n. sp.						*		After P. F. Adams, Surveyor General, N.S.W.,
Caulopteris (?) Adamsi, n. sp.								Newcastle.
III. *Lycopodiaceæ*.								
Lepidodendron nothum, Ung.							*	Goonoo Goonoo and Queensland.
" Australe, M'Coy								Gipps Land.—"different from last named," Feist.

L

APPENDIX XX—continued.

Name.	Uppermost (Jurassic) beds ("Tæniopteris beds"), Tasmania(?), Victoria, & Clarence R.	Wianamatta and Hawkesbury.	(?) Bacchus Marsh Sandstones (Lower Mesozoic in Victoria).	Newcastle beds = Upper Coal Measures in N.S.W. [c]	Lower Coal Measures, Mesozoic flora with Palæozoic fauna. [b]	Strata with Lower Carboniferous flora, Smith's Creek; Port Stephens. [a]	Upper Devonian (?)	Remarks
Cyclostigma Australe, n. sp.							*	Smith's Creek. Lowest Carbonif. (?)Ursa Stufe of Heer.
Lepido. rimosum, Corda						*		Goonoo Goonoo, Queensland.
" dichotomum, Stbn.						*		Rouchel R., N.S.W.
Syringodendron, sp.						*		Rouchel R., N.S.W. } According with quotation by W. B. Clarke.
IV. Cycadaceæ.								
1. Zamiæ.								
Zeugophyllites elongatus, Morr.	*				*?			(?) Jerusalem Basin, Tasmania; Newcastle, according to M'Coy.
Zamites ellipticus, M'Coy	*							Victoria.
" Barklyi,	*							
" longifolius	*			*				Illawarra and Newcastle.
Noeggerathia spathulata, Dan.				*				Newcastle.
" media, Dan.				*				Blackman's Swamp.
" sp.				*				Blackman's Swamp.
(?) Zamia sp.					*			Anvil Creek.
" sp. (quoted by W.B.C.)								
V. Coniferæ.								
Brachyphyllum (?) Australe, n. sp.				**				Blackman's Swamp. Newcastle.
Scales of coniferous plants, Dan.								
Incertæ sedis.								
Cardiocarpum Australe, Carr.	*							Tivoli Mines, Queensland.

a. Strata with plants which by themselves are of *Carboniferous* (partly Lower Carb.) age.

b. Strata the plant remains of which are not by themselves of Carboniferous age, but have to be considered of that age on account of association with Marine fossils of Carboniferous age, as Anvil Creek, Rix's Creek, Greta, &c.

c. See note at end.

Bacchus Marsh Sandstones.—"Now, perhaps, the Bacchus Marsh sandstones have to be placed as above or partly on the horizon of the Newcastle beds. They may, perhaps, also be called the 'Gangamopteris beds'; as so far as I could inform myself, they contain Gangamopteris only (coming in this very close to our Talchír group, underlying the Coal-beds where Gangamopteris preponderates).

The species described from here are:—

   Gangamopteris angustifolia, M'Coy
     " spathulata "
     " obliqua "

'If this position is correctly assigned, then the Indian coal flora (with Glossopteris) is a third re-appearance of the Australian *Lower* flora, in the Upper Marine beds, as in column marked b in the 'Systematic Table.'"

Hawkesbury and Wianamatta Beds.—Dr. Feistmantel makes the following observations respecting these:—"Although representing, perhaps, stratigraphically two groups, they seem to me, from a palæontological point of view, to be of the same age, or very nearly so. It is true they contain fishes, some of which (one, *Palæoniscus*) are heterocercal, but another one seems *not* heterocercal (*Cleitholepis granulatus*)—[But Egerton distinctly calls it heteroc.—W.B.C.]—while of the third one the tail is not known. If we now take into consideration that a *Palæoniscus* is known from the *Karoo* beds in South Africa, which are more than probably Trias, and if we consider that a *Palæoniscus superstes*, Eg., is described from Keuper in England, then we must not be astonished that a *Palæoniscus* should be found in these Hawkesbury and Wianamatta beds, which I would consider *Upper Triassic* (although the plants by themselves would justify to consider these beds as on the horizon of the Rhaetic between Keuper and Lias). These are all I could determine or get information of."

Jurassic or Highest Beds.—Dr. Feistmantel writes:—"There is one point not quite clear to me, if that *Glossopteris* which Professor M'Coy ("*Prodomus*") and Mr. B. Smyth ("*Report of Progress*") mention as occurring in one specimen together with *Pecopteris Australis*, Morr., from Tasmania, belongs also to this group of strata."

General Remarks.—Dr. Feistmantel remarks, with reference to India, "It always results more and more that our Coal-bearing strata are only 'plant-bearing'; no Marine fossils are here. Always more evidence is procured for *Triassic* age of the Coal Measures, and there is no other evidence for the view of their probably Upper Palæozoic age than the generical affinity of some plants, as *Vertebraria, Phyllotheca,* and *Glossopteris*, with the Newcastle beds, and partly in some of your Lower beds. As I have once before mentioned, you may have every reason for an 'Upper Palæozoic' age of your Newcastle beds, but there is nothing of this sort in our Indian Coal-seams.

"As I mentioned also before, our Indian Coal Measures are underlaid by the so called Talchír group, and another group of Coal-seams which I discovered and proved in *two* Coal-basins (in Kurhurbali, in Bengal, and Mohpani, in the Sutpúra Basin), which are characterised by the preponderance of *Gangomopteris*, by the absence (or rareness) of *Vertebraria*, by the rareness of *Glossopteris*, which reminds strongly of the Bacchus Marsh sandstones. If it would be proved that the Bacchus Marsh sandstones are on the horizon of your Newcastle beds or the termination of them, then our Coal flora of India would appear as a *third* repetition of that in Australia (taking the Talchír group as representative of the Bacchus Marsh sandstones, I mean, it would be so):

AUSTRALIA.	INDIA.
	III.—Damúda—Coal-bearing strata, with *Glossopteris, Phyllotheca,* &c.
Bacchus Marsh sandstones, with *Gangamopteris.*	Talchír group, with *Gangamopteris.*
II.—Newcastle beds, with *Glossopteris, Phyllotheca, Vertebraria,* &c.	
Marine Carboniferous animals.	
I.—*Glossopteris, Phyllotheca,* &c., with Marine animals, &c., in New South Wales.	

Dr. Feistmantel writes that he will compare always with the Indian formations his intended description of some of the Australian fossils; and he says, "it will be seen that the Australian Newcastle beds and the Indian Damúdas are not to be confounded."

With reference to the list now published, he says—" I send you to-day again the lists of the plants, &c., based on your first collection [forwarded to Calcutta], and on the quotations by other authors, as I have put it down in my manuscript (it may be, that in your collections, now expected, will be some other forms, which I shall communicate hereafter)."

"The 'Systematic Table' includes those species only which I could determine from your first collection, and which were described before by others, but there is every probability that in your recent collections there will be some other forms."

"Of the two boxes you recently sent me, I received only that entrusted to Professor Liversidge. I found everything in order, and I am very much obliged to you for your great kindness. The specimens of Lepidodendron which were in that box as the *L. nothum* as from Goonoo Goonoo and from Queensland, and the two small specimens from Smith's Creek which you put in extra in an envelope are of great importance. The one (marked 154, Smith's Creek, 1850) is again a *Rhacopteris,* and proves that my determination of the former specimens you sent from Smith's Creek and Port Stephens were correct, when I put them down as *Rhacopteris* (comp.) *inæquilatera,* Göpp; because this new specimen with more split leaves is to the former one (from the same localities) in the same relation as are certain forms in the Kohlenkalk and Culm of *Silesia* to the real *Rhacopteris inæquilatera,* Göpp. I have described these forms from the Silesian Kohlenkalk with more split leaves as Sphenopteris (*Rhacopteris*) *Römeri,* Fstm. (1873, '*Ltschr. d. D. geol. Gesellsh.*') but more material proved hereafter that these very possibly belong as certain developmental states to *Rhacopteris inæquilatera,* being connected with this later species by forms which were described as *Rhac. flabellifera,* Stur. (from the Culm flora).

" There would therefore be the following three forms :—

    (1.) Rhacopteris *inæquilatera*, Göpp.
    (2.)   ,,   *flabellifera*, Stur.
    (3.)   ,,   *Römeri*, Feistm.

Of these forms are present in your Lowest Coal-beds—at *Smith's Creek* the forms Nos. 1 and 3, and at Port Stephens No. 1.

" The other little specimen which you sent from Smith's Creek looks very like some forms which also occur in the Silesian Kohlenkalk, and which I thought could belong to the genus *Psilophyton* or which possibly could perhaps be the fructification of a fern. This will be determined hereafter, but so much is certain that similar forms occur in the Lower Carboniferous."

[Since the foregoing has been in type, I have received a copy of a paper printed in the "*Records of the Geol. Survey of India*," No. 1, 1878, entitled "*The Palæontological relations of the Gondwána System : a Reply to Dr. Feistmantel*, by *W. T. Blanford, F.R.S., Deputy Superintendent, Geol. Survey of India.*" I can now only refer my readers to this document, which continues the discussion as to connection between the Indian and Australian Coal-formations to date.—W.B.C.]

NOTE.—With respect to column headed " Newcastle Beds." It appears to Dr. Feistmantel from the descriptions given, that these beds are above the *Marine* beds, and that they have no Palæozoic fossils in them except the *heterocercal* fish. The flora is completely *Mesozoic*; and considering the position of these beds to the following strata, they should, judged from their flora only, appear best as of Triassic age. But as it seems that stratigraphically (as determined by Mr. Clarke) they are in close connection with the strata in column marked *b*, and as Mr. Clarke insists upon the heterocercal fish (Urosthenes) showing evidence of a Palæozoic age, Dr. Feistmantel thinks there can be no objection to consider these Newcastle beds as terminating the Palæozoic epoch in Australia ; but after which only (he suggests) *Glossopteris* spread out to India and Africa (?), there being NO direct evidence for a Palæozoic age of that epoch during which Glossopteris lived in India—on the contrary, all evidence being in favour of a Triassic epoch. From the first he did NOT consider these beds *younger* than Triassic, but always as Lowest Trias.

Sydney : Thomas Richards, Government Printer.—1878.

SECTION
OF
MOUNT VICTORIA, N.S.W.

Scale, 240 Feet to 1 Inch.

kness Bed.	Nature of Bed.		Height in feet above the Sea.	Remarks.
	Pebbly Sandstones, with bands of hematitic Iron Ore and Veins; and blue and red Shales; Plants.		2525 3505	Summit of Mount Victoria. Level of Shepherd's Toll-bar. Sandstones with Plants, to the E Mount Victoria 1 mile.
165 20 10	Variegated Sandstone, with White and red Shales. Red Shales. Sandstone.		3340 3340 3320	
50 20	Gritty Sandstone.		3230 3260	
70	Sandstones and Grits.		3190	
40	Hard Shales like Silicated Clay of Nobby Island; trace of Coal.		3150	
	Sandstones and Grits.		3000	Level of Coal on descent of Mount and under Hassan's Walls, and Sugar Loaf.
266 75	Indurated Shale, with Glossopteris and thin vein of Coal.		2895 2870	
200	Fine Sandstones; Grits; Silicate of Alumina; and Pebble Bed.		2675 2670	Bed of Reedy Creek [Cannel Coal]. Bottom of Mount Victoria.
100 to 140	Altered Rocks, with Porphyry and Felstone.		2570 2580	Vale of Clwydd.
	Granite.		2304	Lett River at Hartley Bridge.

BUR

	Thickness of Beds.	Nature of Bed.
Hawkesbury Rocks.	780	Yellowish and reddish purple coarse Sandstone abundance of Quartz p and occasional thin se Blue Shale and Ironsto

SECTION
TO ILLUSTRATE THE STRUCTURE OF
BURRAGORANG, NEW SOUTH WALES.

Scale, 240 Feet to 1 Inch.

Nature of Bed.	Height in feet above the Sea.	Remarks
	1996	The summits are much worn and points of rock left in consequence. The whole horizontally bedded, and jointed vertically in three planes, S.W., N.N.W., and S. The upper beds appear to have a varying slope from 12° to N.W. to 10° S.S.W., the effect of local displacements as the horizontal stratification is prominent.
Yellowish and reddish and purple coarse Sandstones, with abundance of Quartz pebbles, and occasional thin seams of Blue Shale and Ironstone.		
	1576	Blue Shale in Eastern Escarpment.
		Blocks fallen from above encumber the whole of the slopes, and obscure the beds of the Coal Measures, rendering detection of seams by the eye almost impossible.
Base of Hawkesbury Rocks.	1245	On West side of Valley, towards head of Lacy's Creek.
Red Shale.	1216	At the base of the Hawkesbury rocks, blocks fallen and accumulated so as to hide the junction. Dense vines, fig-trees, ferns, and jungle, with pools of water.
White Argillaceous Shales and Sandstones.	1120	
Sandstone in Terraces.	1080	
Coal Seam.	937	Level of Fig-tree, a well-known object on descent where the track branches to the Nattai and the Wollondilly.
Shale, Sandstone.	907	Water in springs welling out; temperature of water (20 September, 1855) 50° F., mean temperature at Camp 65° and of Wollondilly 54°.
White and red spotted Shales, with Fossil-wood and Glossopteris.		
Hard white Shale.	758	The surface slope of the Coal Measure beds is about 12°.
Shale, Sandstone, Conglomerate, Ferruginous Sandstone.		
Hard white Shales, with plants.	554	Block of Cannel in dry creek, prismatic, 1′ 9″ thick.
Marine Beds.	537	
	500	Upper observed limit of Muree beds (slope of undercliff 15°) opposite mouth of Nattai.
	468	In dry creek in Carlon's land, Shale with Glossopteris.
Spirifer Sandstone.	410	Under "Tancriffe" with fallen red and white Breccia, capped by quartzose sandstone.
	402	Blocks of Cannel with Glossopteris in dry creek above Chitty's.
Spirifers and Stenopora.	350	
Red and yellow Sandstone.	312	In low ridge on flat, N. 25° W. magnetic.
Brown Shale.	310	Cannel 9 inches thick in loose blocks on flat at Chitty's; waterworn.
Muree Sandstone, with Spirifers and Stems of Plants.	267	Same rock as at Muree and near Worragee on Shoalhaven.
Shale.	240	Level of river banks at camp near mouth of Nattai.
Pebbles of Porphyry and of Coal Shale, Sandstone, &c.	230	
	200	Level of Wollondilly (September 18, 1855) at junction with Nattai. Breadth of the Valley there, about 1½ mile.
Sea Level.		Barometrically deduced in relation to the above heights.

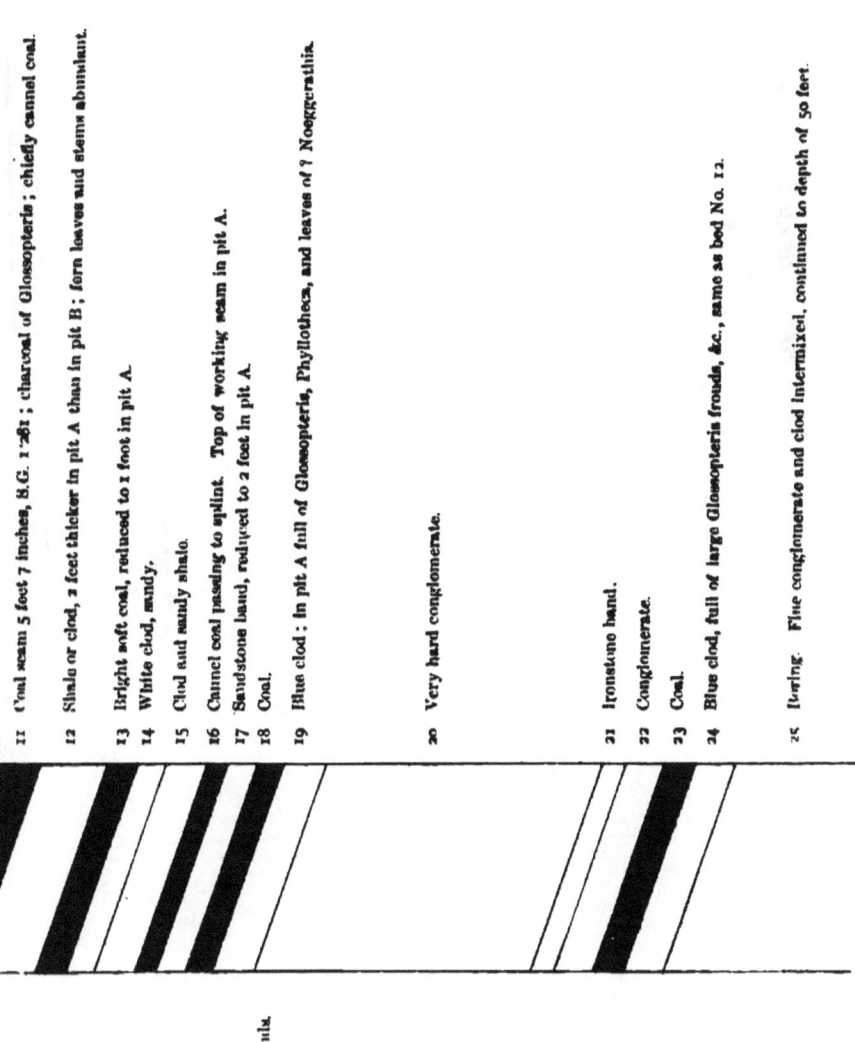

11 Coal seam 5 feet 7 inches, S.G. 1·281; charcoal of Glossopteris; chiefly cannel coal.
12 Shale or clod, 2 feet thicker in pit A than in pit B; fern leaves and stems abundant.
13 Bright soft coal, reduced to 1 foot in pit A.
14 White clod, sandy.
15 Clod and sandy shale.
16 Cannel coal passing to splint. Top of working seam in pit A.
17 Sandstone band, reduced to 2 feet in pit A.
18 Coal.
19 Blue clod: In pit A full of Glossopteris, Phyllotheca, and leaves of ? Noeggerathia.
20 Very hard conglomerate.
21 Ironstone band.
22 Conglomerate.
23 Coal.
24 Blue clod, full of large Glossopteris fronds, &c., same as bed No. 12.
25 Boring. Fine conglomerate and clod intermixed, continued to depth of 50 feet.

Pit B ends.

No. 1.
SECTION OF COAL PITS AT STONY CREEK, N.S. WALES, NEAR WEST MAITLAND.

Dip, E. 6° S. 16°. Scale, 24 feet to 1 inch.

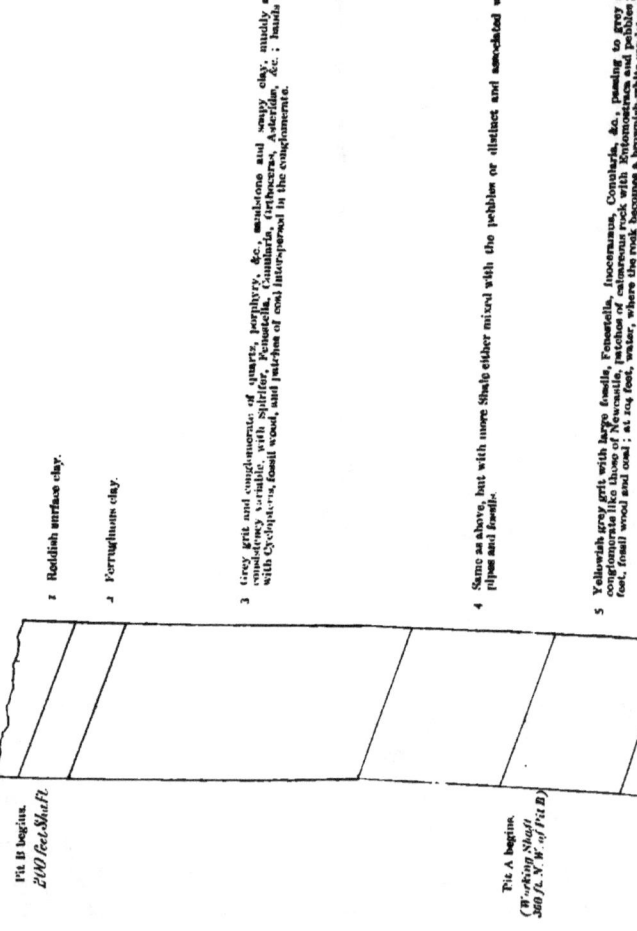

1 Reddish surface clay.

2 Ferruginous clay.

3 Grey grit and conglomerate of quartz, porphyry, &c., sandstone and soapy clay, muddy and blue; consistency variable, with Spirifer, Pentaclla, Cassidaria, Orthoceras, Asteridæ, &c.; bands of Shale, with Cyclopteris, fossil wood, and patches of coal interspersed in the conglomerate.

4 Same as above, but with more Shale either mixed with the pebbles or distinct and associated with coal pipes and fossils.

5 Yellowish grey grit with large fossils, Fenestella, Inoceramus, Conularia, &c., passing to grey grit and conglomerate like those of Newcastle, patches of calcareous rock with Entomostracea and pebbles; at 100 feet, fossil wood and coal; at 104 feet, water, where the rock becomes a brownish-white substance.

Pit B begins. 200 feet Shaft.

Pit A begins. (Working Shaft) 360 ft. N.W. of Pit B.

6 Conglomerate; with vegetable impressions in light blue shale amidst it and a few Spirifera.

7 Rotten, soft, decomposing conglomerate, of a greenish hue, with Spirifer, coal pipes, shale, fossil wood, and Ironstone. Shaft obliged to be timbered.

8 Very hard, fine conglomerate, with coal pipes, passing to grit.

9 Coarse conglomerate, with pebbles of flinty rock and unrolled fragments of sandstone.

10 Ditto with well rounded pebbles; slightly effervescent.

11 Coal seam 5 feet 7 inches, S.G. 1·281; charcoal of Glossopteris; chiefly cannel coal.

12 Shale or clod, 3 feet thicker in pit A than in pit B; fern leaves and stems abundant.

13 Bright soft coal, reduced to 1 foot in pit A.

14 White clod, sandy.

15 Clod and sandy shale.

16 Cannel coal passing to splint. Top of working seam in pit A.

17 Sandstone band, reduced to 2 feet in pit A.

18 Coal.

19 Blue clod; in pit A full of Glossopteris, Phyllotheca, and leaves of ? Noeggerathia.

20 Very hard conglomerate.

21 Ironstone band.

22 Conglomerate.

23 Coal.

24 Blue clod, full of large Glossopteris fronds, &c., same as bed No. 12.

Fig. 1.

SECTION OF COAL PITS AT STONY CREEK, N.S. WALES, NEAR WEST MAITLAND.

Scale 150 × 60 × 3 feet. 1 inch to 1 foot.

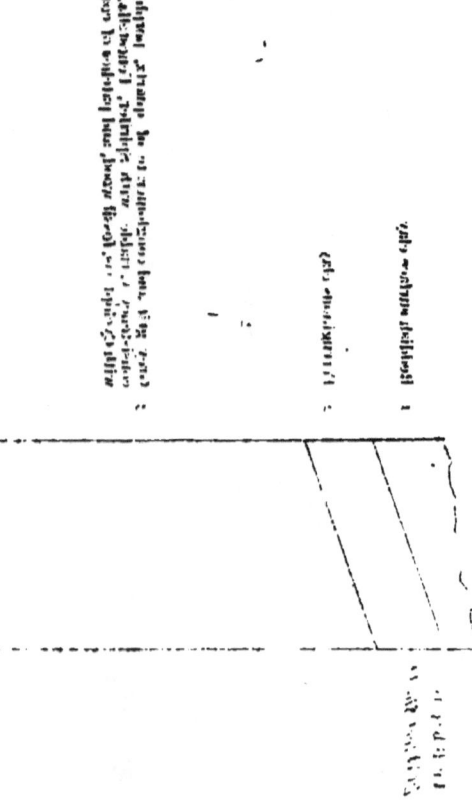

Reddish yellow clay

1/4 indurated clay

Laminated clay
Deep clay

Reddish yellow clay, arenaceous, &c., rhythmic change of transparent and light blue; indurated clay, argillites, tuffaceous alternating with Shale of slaty development in this conglomerate.

No.	Description	ft	in
14	Coarse Grey Sandstone (Thin Layers) in Conglomerate*	18	
15	Conglomerate	6	3
16	Mild Blue Stone*	4	6
17	Conglomerate, Marine Shells	5	
18	Grey Sandstone	7	
19	Conglomerate	4	3
20	Blue Shale	1	6
21	Conglomerate	1	
22	Black Stone .... Coal	4	
	Clay Bands .... Coal	8	
	Fire Clay .... Coal	3	
	Black Sandstone .... Coal	12	5
23	Blue Shale	4	6
	Total	4 37	0½

14. Inoceramus; Conularia; Spirifer *glaber*; patches of wood and Inoceramus shell.
15. Black and green pebbles; decomposing shale, &c.
16. Semi-conchoidal fracture and brittle.
17. Coarser than No. 15. Pyrites on pebbles; Sp. *glaber*; Conularia; and fragments of shells.
18. Fragments of rocks and shells.
19. Dark grey coarse base; smooth pebbles, and Sp. *Tasmaniensis*, with carbonized wood.
20. Glossopteris and allied genera, Pyrites.
21. Large pebbles, cubic pyrites, wood and coal.
21 B. Light brown flaky rock, with numerous root-like fragments; surface coaly ["black stone."]
Fragments of ferns and light coal seam.

* The rocks of Nos. 5, 14, 18, effervesce with acid

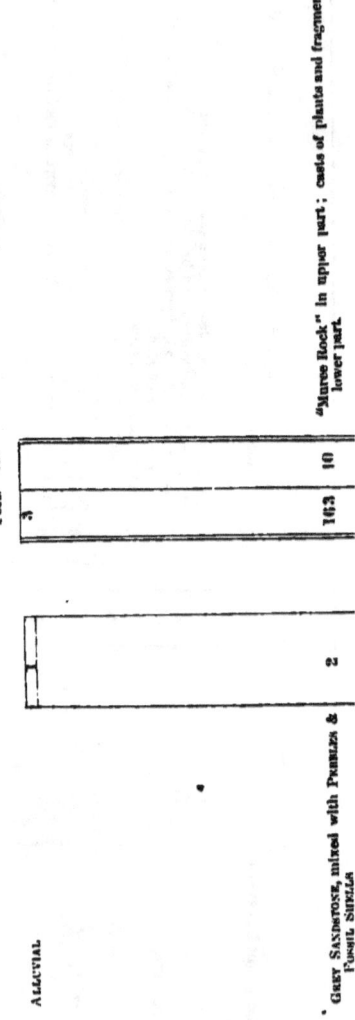

Layer	Rock	ft	in	Description
6	DARK GREY SANDSTONE	42	0	Concretionary spherical nodules, as at Wollongong, Jervis Bay, &c.
7	BLUE METAL	12		Elliptical nodules and pebbles.
8	CONGLOMERATE	4		Large pebbles; Conularia and Spirifers of various species.
9	DARK GREY SANDSTONE	9		Small pebbles.
10	CONGLOMERATE	2		Hard, grey, and full of Spirifers.
11	BLUE METAL, with Pebbles	25		Spirifers; flinty shale, and junks of wood with calcareous rings
12	CONGLOMERATE	8		Dark gritty, with Spirifer fragments and carbonized wood.
13	Blue Marls	26		Concretions; wood and pebbles.
14	COARSE GREY SANDSTONE (THIN LAYERS) in CONGLOMERATE *	18		Inoceramus; Conularia; Spirifer *glaber*; patches of wood and Inoceramus shell.
15	CONGLOMERATE	6	3	Black and green pebbles; decomposing shale, &c.
16	MILD BLUE STONE *	4	4	Semi-conchoidal fracture and brittle.
17	CONGLOMERATE, MARINE SHELLS	5	7	Coarser than No. 15, Pyrites on pebbles; Sp. *glaber*; Conularia; and fragments of shells.
18	GREY SANDSTONE	7	4	Fragments of rocks and shells.
19	CONGLOMERATE	4	1½	Dark grey coarse base; smooth pebbles, and Sp. *Tasmaniensis*, with carbonized wood.
20	BLUE SHALE, 2'0 ... Coal	1	7	20. Glossopteris and allied genera, Pyrites.
21	BLACK STONE Conglomerate, 1'0 ... Coal	4		21. Large pebbles, cubic pyrites, wood and coal.
	FIRE CLAY, CLAY BANDS ... Coal	8		21 B. Light brown flaky rock, with numerous root-like fragments; surface easily ["black stone."]
	BLACK SANDSTONE ... Coal	3		
23	BLUE SHALE	12		Fragments of ferns and light coal seam.
	Coal	4	6	
	Total	**4.37**	**0½**	

*The rocks of Nos. 5, 14, 16, effervesce with acid

www.ingramcontent.com/pod-product-compliance
Lightning Source LLC
Chambersburg PA
CBHW020248170426
43202CB00008B/279